NUCLEOSYNTHESIS

NUCLEOSYNTHESIS
Challenges and
New Developments

Edited by
W. David Arnett
and
James W. Truran

The University of Chicago Press

Chicago & London

W. David Arnett is professor in the Department of Physics, the Department of Astronomy and Astrophysics, and the Enrico Fermi Institute at the University of Chicago. James W. Truran is professor of astronomy at the University of Illinois.

The University of Chicago Press, Chicago 60637
The University of Chicago Press, Ltd., London

© 1985 by The University of Chicago
All rights reserved. Published 1985
Printed in the United States of America

94 93 92 91 90 89 88 87 86 85 5 4 3 2 1

Library of Congress Cataloging in Publication Data

Main entry under title;

Nucleosynthesis: challenges and new developments.

Papers from a symposium held at Yerkes Observatory,
University of Chicago, 1983.
    Bibliography: p.
    1. Nucleosynthesis—Congresses. I. Arnett, W.
David (William David), 1940–   . II. Truran,
James W.
QB450.N8   1985   523.8   85-1160
ISBN 0-226-02787-2
ISBN 0-226-02788-0 (pbk.)

Dedicated to W. A. Fowler on the announcement of the Nobel Prize for Physics.

Contents

Preface

A script so written would be thought implausible. Yet it did happen. At 6 a.m., three hours before he was to give the first talk of the conference, word spread that William A. Fowler would share the 1984 Nobel Prize in physics with S. Chandrasekhar. After an early morning telephone barrage, Willy walked into the hotel breakfast room to the applause of the astrophysicists and the quizzical looks of the others. The hotel operator "just could not understand" who were these people who elicited such interest from the international press. Despite the camera crews, who overdosed on shots of the Yerkes dome and the Wisconsin autumn leaves, the conference proceeded with a strange sort of buoyant happiness that came from the opportunity to share with Willy the joy of the honor.

Willy's prize was clearly the highlight of the conference. It did not obscure the substantial contributions which follow. Not all of the talks given are represented here. Excellent talks were presented by J. Gallagher ("Simple Tests for Chemical Pollution Processes in Galaxies"), T. Weaver ("Massive Star Evolution"), and S. E. Woosley ("Type I Supernovae"). We are pleased that so many of the authors were able to take time to help provide a record of this unique conference.

We wish to thank Jennie Lightner and Susan Lancaster for their willingness to become so expert with computer typesetting, and to use their skill here.

W. D. Arnett
Chicago

J. W. Truran
Urbana

1

Introduction

W. David Arnett and James W. Truran

Nucleosynthesis theory has progressed significantly over the past quarter century. The seminal papers are by Burbidge et al. (1957) and Cameron (1957). Stars and supernovae within our own Galaxy have been determined to be the most important nucleosynthesis sites. The specific nuclear processes which play a role in the synthesis of heavy nuclei have been identified. Several factors were critical in stimulating these rapid developments. Improved determinations of element abundances in diverse astronomical objects have provided clues to and imposed boundary conditions upon the transformations by which nuclei heavier than hydrogen and helium are synthesized in stars and supernovae. Advances in nuclear physics allowed better experimental and theoretical determinations of the thermonuclear reaction and weak interaction rates which are relevant to nuclear energy generation and nucleosynthesis. Advances in computer technology have provided successively better generations of computers which now allow detailed calculations of nucleosynthesis to be carried out in the context of increasingly realistic models of the relevant environments. Recent global reviews of nucleosynthesis theory include those by Trimble (1975) and Truran (1984). There are a number of more specialized reviews of explosive charged particle nucleosynthesis (Arnett 1973; Truran 1973), of mechanisms of neutron capture synthesis (Hillebrandt 1978; Truran 1980; Ulrich 1982), and of the origin of the light elements (Reeves 1974; Audouze and Reeves 1982).

The aim of this meeting was to call attention to the significant problems and challenges in nucleosynthesis research. We begin with reviews of some recent results relating to nuclear cross-section data, which provide the empirical basis for nucleosynthesis theories. This is followed by a review of cosmological nucleosynthesis, then galactic evolution, on to the implications of cosmic rays for interstellar medium isotopic abundances, and to isotopic anomalies in the meteorites. Having sketched the broad picture, we examine stellar nucleosynthesis in detail, beginning with the more massive stars, core collapse, and type II supernovae and proceeding to type I supernovae and intermediate mass stars and, finally, to novae, the least massive stars that are discussed.

Nucleosynthesis studies necessarily require, as input, a substantial variety of nuclear parameters. Both thermonuclear reaction rates and weak interaction rates are needed. Experimental information is available for a number of reactions of importance, particularly for hydrogen and helium burning which occur most of the lifetime of a star. However, it clearly is not feasible to obtain direct information for each of the hundreds of nuclear reactions which may be involved in the later stages of nuclear energy generation in stellar interiors or in nucleosynthesis processes associated with stars and supernovae. It is necessary to develop models of nuclear reactions and of weak interactions to the point where reliable theoretical predictions become possible. Even so, experimental data are critical for a reliable systematic treatment. A brief review of the current theoretical estimates of thermonuclear reaction and weak interaction rates is provided by Truran (1984).

Several of the papers in these proceedings address questions concerning thermonuclear reaction rates. Willy Fowler ("Back to Square One in Nuclear Reaction Rates") reviews the current situation with regard to critical cross sections of reactions in the proton-proton hydrogen burning sequences and the $^{12}C(\alpha,\gamma)$ ^{16}O reaction which participates in helium burning. Claus Rolfs ("Status of Helium Burning of ^{12}C") further elaborates the experimental situation with regard to the $^{12}C(\alpha,\gamma)$ ^{16}O reaction. Experimental and theoretical determinations of the reduced alpha width of the 7.12 MeV excited state of ^{16}O span a considerable range, within which limits the helium burning abundances may vary dramatically. The net production of ^{12}C and ^{16}O, integrated over contributions from massive stars and red giants, is roughly compatible with the solar system C/O value of 0.6, but specific questions regarding C/O ratios in massive stars and supernova cores yield critical uncertainties in predictions of supernova nucleosynthesis. Studies of the neutron capture process, which are responsible for the bulk of the nuclear species more massive than iron, involve the use of vast networks of nuclear reactions. Franz Käppeler ("Precise Stellar Neutron Capture Rates: s-Process Needs and

Experimental Possibilities") has reviewed the experimental situation regarding neutron capture cross sections, emphasizing those which are critical to studies of *s*-process nucleosynthesis.

The major boundary conditions within which we seek to form theories of nucleosynthesis are provided by determinations of abundances in diverse astronomical environments. Important recent compilations of solar system abundances of the elements have been provided by Anders and Ebihara (1982) and Cameron (1982). Reviews of our current knowledge of the elemental and isotopic compositions of the sources of the galactic cosmic rays include those by Mewaldt (1983) and Simpson (1983). Abundances in the stars and gas in our Galaxy and other galaxies have been reviewed by a number of authors (see, for example, the references given by Trimble (1975) and Truran (1984)). Among the areas discussed in these proceedings are isotopic abundances in meteorites, in the cosmic rays, and abundances in supernova ejecta. These abundance determinations may constitute direct tests of nucleosynthesis.

Advances in computer technology have made possible increasingly better numerical calculations of the structure and evolution of the astrophysical objects in which nucleosynthesis is believed to occur. Active research tends to be concerned with the operation of specific mechanisms of nucleosynthesis in the context of realistic models of these environments. This is reflected in many papers in these proceedings. Gary Steigman ("Primordial Nucleosynthesis: Looking Good!") critically compares theory and observation with respect to the standard model of big bang nucleosynthesis. He finds complete consistency between the abundances of the light elements D, ^{3}He, and ^{7}Li predicted by primordial nucleosynthesis in the standard, hot big bang model and the primordial abundances of those elements inferred from current data.

Nucleosynthesis theory is intimately related to galactic evolution, which itself is imperfectly understood (to say the least). Don Clayton ("Galactic Chemical Evolution and Nucleocosmochronology: A Standard Model") presents a new and elegant analytic tool for a more realistic exploration of this problem and its connection to cosmochronology. Cosmic ray isotopic abundance ratios for the important elements neon, magnesium, and silicon are currently thought to reflect the interstellar medium now, while the meteoritic values are generally thought to refer to the value when the sun was formed

4.5 billion years ago. Mark Wiedenbeck ("Cosmic Rays as a Possible Probe of the Composition of Interstellar Matter") reviews the measurements and their significance.

Recent studies of isotopic abundances in meteoritic material have revealed the presence of nonsolar patterns for a number of elements (O, Ne, Mg, Si, Ca, Kr, Sr, Xe, Ba, Nd, and Sm) and show that short-lived radioactive nuclei (^{26}Al, ^{107}Pd, ^{129}I, and ^{244}Pu) were present in the early solar system. These abundance anomalies and their possible interpretations have previously been reviewed by a number of authors (Clayton 1978; Podosek 1978; Lee 1979; Wasserburg, Papanastassiou, and Lee 1980; Wasserburg and Papanastassiou 1982). Dave Schramm ("Nucleosynthetic Interpretations of Isotopic Anomalies") discusses possible interpretations of isotopic anomalies, in the context of nucleosynthesis theory, emphasizing the critical role these studies play in increasing our understanding of stellar and galactic evolution and nucleosynthesis.

Detailed models of the evolution of massive stars to their pre-supernova configuration and hydrodynamic calculations of supernova events provide essential input to nucleosynthesis studies: predictions of the temperature and density history of the stellar and supernova-ejected matter. Dave Arnett ("Late Stages of the Evolution of Massive Stars: Challenges") reviews the current theoretical and observational uncertainties with regard to stellar evolution, with emphasis on the massive stars which constitute the likely progenitors of type II supernovae. Wolfgang Hillebrandt and Rudiger Wolff ("Models of Type II Supernova Explosions") then present the results of recent calculations of the hydrodynamic evolution of type II supernovae. Nucleosynthesis considerations (production of the elements from silicon to iron and nickel) favor the view that at least some massive stars explode by the core-bounce mechanism, though hydrodynamic studies to date have not provided a definitive statement on this matter. Their results confirm that a small iron-core mass and low initial entropy both act to favor an explosion. Friedrich Thielemann and Dave Arnett ("Hydrostatic Nucleosynthesis and Pre-Supernova Abundance Yields of Massive Stars") present the results of detailed calculations of nucleosynthesis accompanying hydrostatic phases of core helium, carbon, neon, oxygen, and silicon burning in stars with helium core masses ranging from 1.5 to 16 $M_\odot$, showing the effects of the uncertain ^{12}C(α,γ) ^{16}O reaction. In particular, it appears that varying the ^{12}C(α,γ) ^{16}O rate within the experimental uncertainties will remove problems with recent estimates of nucleosynthesis yields from massive stars. The pervasive role of supernovae in heavy element synthesis is suggested by several factors: extreme temperature and density conditions allow the complex nuclear processes to occur, evolution on a hydrodynamic time scale is compatible with inferred nuclear time scales, and nuclear processes occur with expansion, cooling, and mass ejection, thereby ensuring that the abundance patterns reach the interstellar medium unscathed. Al Cameron, John Cowan, and Jim Truran ("Neutron Source Requirements for the Helium r-Process") discuss the restrictions their calculations place on the operation of the r-process under explosive conditions in supernova environments. Their studies strongly suggest that the astrophysical site appropriate to r-process neutron-capture nucleosynthesis has not yet been clearly identified.

Observations of supernova ejecta (Chevalier and Kirshner 1979; Branch et al. 1981; Shull 1982; Branch et al. 1983; Branch 1984) reveal enrichments of some heavy element abundances relative to solar abundances, confirming that significant nuclear processing of this material has indeed occurred. This appears to be the most direct probe of the supernova nucleosynthesis conditions. To give a basis for the interpretation of each supernova spectra, Dave Branch ("The Optical Spectrum of a Carbon-Deflagration Supernova") has computed theoretical optical spectra for a specific hydrodynamical supernova model: the carbon deflagration of an accreting carbon-oxygen white dwarf. The theoretical spectra agree with spectra of the type I supernova 1981b in NGC 4536 on a number of important features. Ken'ichi Nomoto ("Explosive Nucleosynthesis in Carbon Deflagration Models for Type I Supernovae") presents detailed spherically symmetric calculations of explosive charged-particle nucleosynthesis occurring in carbon deflagration models of accreting white dwarfs. Explosive nucleosynthesis in the deflagration wave both produces sufficient ^{56}Ni to power the light curve of SN I by radioactive decay and synthesizes significant amounts of the intermediate mass nuclei O, Mg, Si, S, Ar, and Ca. Ewald Müller and Dave Arnett ("Carbon Combustion Models of Type I Supernovae: The Propagation of the Thermonuclear Burning Front") address a question crucial to models of type I supernova explosions. Their numerical studies of the propagation of thermonuclear burning fronts, which allow for the presence of two-dimensional hydrodynamic flows, indicate that, in almost all cases examined, a spherical burning front evolves to a nonspherical deflagration of a degenerate carbon-oxygen core. It appears that this affects the nucleosynthesis predicted.

Nucleosynthetic contributions from other astronomical sites may be important. Icko Iben ("Nucleosynthesis in Low and Intermediate Mass Stars on the Asymptotic Giant Branch") considers single stars in the mass range $1 \lesssim M \lesssim 9 \, M_\odot$ and reviews their complex evolution along the asymptotic giant branch and the associated nuclear processes which produce the bulk of the s-process heavy elements. Jim Truran ("Nucleosynthesis in Novae") discusses the theoretical and observational bases for predictions of nucleosynthesis accompanying nova explosions. Relatively low amounts of matter are processed through novae, suggesting that they are not major contributors to the abundances of the bulk of the heavy elements. They may, however, produce important amounts of the rare CNO isotopes ^{13}C, ^{15}N, and ^{17}O, as well as provide a site for the production of both ^{26}Al and a mass 22 enriched neon component. With mention of ^{26}Al we have come almost a full turn, connecting back to the papers of Clayton and Schramm in particular.

We hope that these proceedings will acquaint the reader with current problems, current efforts, and the excitement of this area of research.

References

Anders, E., and Ebihara, M. 1982, *Geochim. Cosmochim. Acta,* **46**, 2363.

Arnett, W. D. 1973, *Ann. Rev. Astr. Ap.,* **11**, 73.

Audouze, J., and Reeves, H. 1982, in *Essays in Nuclear Astrophysics,* ed. C. A. Barnes, D. D. Clayton, and D. N. Schramm, p. 355 (Cambridge: Cambridge University Press).

Branch, D. 1984, in *Stellar Nucleosynthesis,* ed. C. Chiosi and A. Renzini (Dordrecht: Reidel).

Branch, D.; Falk, S. W.; McCall, M. L.; Rybski, P.; Nomoto, A. K.; and Wills, B. J. 1981, *Ap. J.,* **244**, 780.

Branch, D.; Lacy, C. H.; McCall, M. L.; Southerland, P. G.; Uomoto, A.; Wheeler, J. C.; and Willis, B. J. 1983, *Ap. J.,* **270**, 123.

Burbidge, E. M.; Burbidge, G. R.; Fowler, W. A.; and Hoyle, F. 1957, *Rev. Mod. Phys.,* **29**, 547.

Cameron, A. G. W. 1957, Atomic Energy of Canada Ltd., CRL-41.

——. 1982, in *Essays in Nuclear Astrophysics,* ed. C. A. Barnes, D. D. Clayton, and D. N. Schramm, p. 23 (Cambridge: Cambridge University Press).

Chevalier, R. A., and Kirshner, R. P. 1979, *Ap. J.*, **233**, 154.

Clayton, R. N. 1978, *Ann. Rev. Nucl. Sci.*, **28**, 501.

Hillebrandt, W. 1978, *Space Sci. Rev.*, **21**, 639.

Lee, T. 1979, *Rev. Geophys. Space Phys.*, **17**, 1591.

Mewaldt, R. A. 1983, *Rev. Geophys. Space Phys.*, **21**, 295.

Podosek, F. A. 1978, *Ann. Rev. Astr. Ap.*, **16**, 293.

Reeves, H. 1974, *Ann. Rev. Astr. Ap.*, **12**, 437.

Shull, J. M. 1982, *Ap. J.*, **262**, 308.

Simpson, J. A. 1983, *Ann. Rev. Nucl. Part. Sci.*, **33**, 323.

Trimble, V. 1975, *Rev. Mod. Phys.*, **47**, 877.

Truran, J. W. 1973, *Space Sci. Rev.*, **15**, 23.

———. 1980, *Nukleonika*, **25**, 1463.

———. 1984, *Ann. Rev. Nucl. Part. Sci.*, **34**, 53.

Ulrich, R. K. 1982, in *Essays in Nuclear Astrophysics*, ed. C. A. Barnes, D. D. Clayton, and D. N. Schramm, p. 301 (Cambridge: Cambridge University Press).

Wasserburg, G. J., and Papanastassiou, D. A. 1982, in *Essays in Nuclear Astrophysics*, ed. C. A. Barnes, D. D. Clayton, and D. N. Schramm, p. 77 (Cambridge: Cambridge University Press).

Wasserburg, G. J., Papanastassiou, D. A., and Lee, T. 1980, in *Early Solar System Processes and the Present Solar System*, p. 144 (Bologna: Soc. Italiana di Fisica).

Back to Square One in Nuclear Reaction Rates

William A. Fowler

2.1 Introduction

The last several years have witnessed exciting new developments in and lively controversy concerning certain cross sections of reactions in the pp-chain of hydrogen burning and the $^{12}C(\alpha,\gamma)^{16}O$ reaction of helium burning. Since H and He burning are the first two nuclear stages in stellar evolution, the reason for my choice of title will I hope be obvious. Some changes in the thermonuclear reaction rates published in Harris et al. (1983) must be made, and these will be incorporated in a forthcoming publication by Caughlan et al. (1984). It seemed appropriate to present these changes to potential users attending this conference.

2.2 The pp-Chain

The rates of the $^{3}He(\alpha,\gamma)^{7}Be$ and the $^{7}Be(p,\gamma)^{8}Be$ are of great interest in connection with the solar neutrino problem because these reactions lead to energetic neutrinos from $^{7}Be(e^{-},\nu)^{7}Li$ and $^{8}B(e^{+}\nu)^{8}Be(\alpha)^{4}He$. These energetic neutrinos are detectable by the $^{37}Cl/^{37}Ar$ technique employed by Davis, Cleveland, and Rowley (1983) to measure the solar neutrino flux. The flux expected in the Davis observations is almost but not exactly linearly dependent on the rates in the Sun of the two reactions under discussion.

Measurements of the zero energy cross-section factor for $^{3}He(\alpha,\gamma)^{7}Be$ extending over the last two decades are shown in figure 2.1. It will be noted that the Münster results of Kräwinkel et al. (1982) differ by several standard deviations from the early Kellogg/Caltech measurements by Parker and Kavanagh (1963) and Nagatani et al. (1969). This difference precipitated a number of new measurements by Osborne et al. (1982) at Kellogg, who

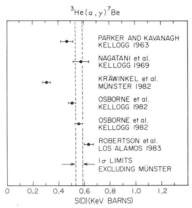

Fig. 2.1. Measurements of the zero energy cross-section factor for $^3\text{He}(\alpha,\gamma)^7\text{Be}$.

detected both the gamma-rays and the ^7Be-decays from $^3\text{He}(\alpha,\gamma)^7\text{Be}$, and by Robertson et al. (1983) at Los Alamos, who detected the ^7Be-decays. A mean value of the measurements in figure 2.1 with a reasonable 1 σ standard deviation can be obtained if the Münster value is excluded. This value is 0.558 $\pm$ 0.037 KeV-barns compared to the Münster value of 0.30 $\pm$ 0.03 KeV-barns and is the value used to determine the reaction rate given in Harris et al. (1983). Those who prefer the Münster value need only decrease the numerical coefficient of the analytical expression for the $^3\text{He}(\alpha,\gamma)^7\text{Be}$ rate given in Harris et al. (1983) from 5.79×10^6 to 3.11×10^6.

As noted previously the decay of ^7Be is frequently used in the measurement of the cross section for $^3\text{He}(\alpha,\gamma)^7\text{Be}$. Electron capture by ^7Be leads to the first excited state of ^7Li, which decays by emission of an easily detectable 0.478 MeV gamma-ray. The branching ratio to this state must be known from independent measurements, and these are summarized without detailed elaboration in table 2.1, which is taken from Skelton and Kavanagh (1984). The Münster results of Trautvetter et al. (1983) can be seen to differ by a number of standard deviations from the weighted mean of the pre-1983 and other 1983 results and to be roughly 50% higher. It will be noted that the use of the Münster branching ratio would bring the results of Robertson et al. (1983) for the $^3\text{He}(\alpha,\gamma)^7\text{Be}$ reaction into fair agreement with those of Kräwinkel et al. (1982). In any case, laboratories in Canada and the United States rallied round and showed that the Münster results were completely in error (see table 2.1). Skelton and Kavanagh (1984) give detailed references.

Table 2.1

1983 Measurements of the ^{7}Be Branching Ratio B_γ
to the First Excited State of ^{7}Li (0.478 MeV Gamma-Ray)

Author	Institution	B_γ
Trautvetter et al.	I. Kernph. Münster	15.4 ± 0.8
Norman et al.	U. Washington	9.8 ± 0.5
Balamuth et al.	U. Pennsylvania	10.10 ± 0.45
Donoghue et al.	Ohio State	10.6 ± 0.3
Davids et al.	Argonne N. L.	10.61 ± 0.23
Fisher and Hershberger	U. Kentucky	10.61 ± 0.17
Mathews et al.	Livermore N. L.	10.7 ± 0.2
Taddeuci et al.	Ohio U.	10.3 ± 1.1
Knapp et al.	Princeton U.	10.9 ± 1.1
Evans et al.	Queens U.	11.4 ± 0.7
Skelton and Kavanagh	Caltech	10.49 ± 0.07
Weighted mean excluding Münster result		10.52 ± 0.06
Weighted mean of pre-1983 results		10.38 ± 0.06
Weighted mean (pre-1983 & 1983)		10.45 ± 0.04

Note: Full references are given in Skelton and Kavanagh (1984).

The most careful measurements were made in Kellogg/Caltech, as shown by the standard deviations that are listed in table 2.1. The Münster results were withdrawn.

The situation in regard to measurements of the S-factor for ^{7}Be$(p,\gamma)^8$B are much more harmonious as indicated in table 2.2. The beautiful recent work of Filippone et al. (1983) is in substantial agreement with previous work in Kellogg/Caltech and elsewhere, but has smaller standard deviations. Filippone et al. (1983) give a weighted average for the zero energy S-factor for ^{7}Be$(p,\gamma)^8$B equal to 0.0238 ± 0.0023 KeV-barns as shown in table 2.2. This value has been accepted by Caughlan et al. (1984) and changes the coefficients of the two terms in the analytical expression for the rate of ^{7}Be$(p,\gamma)^8$B in Harris et al. (1983) from 4.02×10^5 to 3.09×10^5 and 3.30×10^3 to 2.53×10^3, respectively.

The effect of all of these changes on the theoretical expectations for the solar neutrino flux in solar neutrino units (1 SNU $= 10^{-36}$ neutrino captures

Table 2.2

Zero Energy S-Factor for $^7Be(p,\gamma)^8B$

Reference	$S_{17}(0)$ (keV-barns)
Kavanagh (1960)	0.016 ± 0.006 *
Parker (1966)	0.028 ± 0.003 *
Kavanagh et al. (1969)	0.0273 ± 0.0024 *
Vaughn et al. (1970)	0.0214 ± 0.0022 *
Wiezorek et al. (1977)	0.045 ± 0.011 [†]
Filippone et al. (1983)	0.0221 ± 0.0028 *
	0.0206 ± 0.0030 [†]
Weighted average	0.0238 ± 0.0023
Bahcall et al. (1982)	0.029 ± 0.010

Note: Full references are given in Filippone, Elwyn, and Davids (1983).
* ^{7}Be areal density determined from $\sigma(0.77$ MeV) $=$ 157 mb for $^7Li(d,p)^8Li$ (^{7}Li from ^{7}Be-decay).
[†]^{7}Be areal density determined from gamma-ray activity.

s^{-1} per target nucleus) is shown in table 2.3. Careful perusal of table 2.3 will convince even the skeptic that the solar neutrino problem is still with us. This constitutes a basic challenge to neutrino physics, and the theory of stellar structure and evolution. The nuclear reaction rates are now based on careful measurements in a number of laboratories. It is generally agreed that observations of the approximately solar-model independent flux of the pp-neutrinos from the first reaction of the pp-chain must now be attempted. $^{71}Ga/^{71}Ge$ is the preferred technique. The $^{37}Cl/^{37}Ar$ detector must be kept in operation until it can be operated simultaneously with a new $^{71}Ga/^{71}Ge$ detector either in the United States or abroad.

2.3 The $^{12}C(\alpha,\gamma)^{16}O$ Reaction

Since $^{16}O(\alpha,\gamma)^{20}Ne$ has a relatively low reaction rate, helium burning consists primarily of two reactions. The rate for the first of these, $3\alpha \rightarrow {}^{12}C$, is generally considered to be well established although special treatment of the reaction rate is necessary at low temperature as shown many years ago by Cameron (1954) and as emphasized recently by Nomoto (1982). Thielemann

Table 2.3
Solar Neutrino Yields in SNU (10^{-36} s^{-1} per ^{37}Cl target)
Theoretical SNU Calculations: Bahcall et al. (1982)
(All 1 σ Errors)

^{3}He(α,γ)^{7}Be Rate	SNU
CALTECH (Osborne et al. 1982)	7.6 ± 1.1*
$S_{34} = 0.52 \pm 0.03$ keV-b (γ & ^{7}Be-decay)	6.9 ± 1.0†
MUNSTER (Kräwinkel et al. 1982)	5.0 ± 0.7*
$S_{34} = 0.30 \pm 0.03$ keV-b (γ only)	
LANL & MSU (Robertson et al. 1982)	8.8 ± 1.3*
$S_{34} = 0.63 \pm 0.04$ keV-b (^{7}Be-decay only)	
OBSERVED (Davis et al. 1983)	1.8 ± 0.3
REQUIRES:	
$S_{34} = 0.09$ keV-b $\begin{cases} 7\ \sigma\ \text{from Münster} \\ 14\ \sigma\ \text{from Caltech} \\ 14\ \sigma\ \text{from LANL/MSU} \end{cases}$	

*Used 1 σ (S_{34}) = ± 0.05 keV-b. Decrease by $\sim$10% for ± 0.03 keV-b.
†Using $\bar{S}_{17} = 0.024 \pm 0.002$ keV-b, Filippone et al. (1983). (0.029 $\pm$ 0.003 keV-b, Bahcall et al. 1982)

(1983) has provided a general solution of the problem which will be incorporated in Caughlan et al. (1984).

The rate for the second of the important helium-burning reactions, ^{12}C(α,γ)^{16}O, has recently come into question for both experimental and theoretical reasons. Recent measurements by Kettner et al. (1982) at Münster yield somewhat larger cross sections than the Kellogg measurements of Dyer and Barnes (1974). New attempts to obtain data at lower energies than heretofore ($\sim$1.4 MeV) are now under way in both laboratories. Langanke and Koonin (1984) have analyzed the experimental data to determine the astrophysical S-factor. The absolute E2 capture cross section was calculated in a microscopically based direct-capture model, while the E1 cross section was described by the hybrid R-matrix model. Their extrapolation of the data of Dyer and Barnes will be included in Caughlan et al. (1984) with the changes from Fowler, Caughlan, and Zimmerman (1975) now to be noted. The coefficient of the first term (E1) in the analytical expressions for the $^{12}C(\alpha,\gamma)^{16}O$ reaction rate is increased from 9.03×10^7 to 2.93×10^8. The coefficient of the second term (E2) is increased from 2.74×10^7 to 3.14×10^8. This results, for example, in an increase by a factor of ~ 3 at the effective energy, 0.3 MeV, corresponding to a temperature equal to 1.8×10^8 K, a representative helium-burning value. It must be emphasized that these changes are not to be taken as final. New measurements at lower energies will certainly improve and may change the extrapolation to effective helium-burning energies. If users find that their results in a given study are sensitive to the rate of the $^{12}C(\alpha,\gamma)^{16}O$ reaction, then they should repeat their calculations with 0.5 times and 2 times the values recommended here.

The moral of this article is that there are both "certainties" and uncertainties in thermonuclear reaction rates of astrophysical interest. Continued experimental and theoretical studies are in order.

References

Bahcall, J. N.; Huebner, W. F.; Lubow, S. H.; Parker, P. D.; and Ulrich, R. K. 1982, *Rev. Mod. Phys.*, **54**, 767.

Cameron, A. G. W. 1959, *Ap. J.*, **130**, 916.

Caughlan, G. R.; Fowler, W. A.; Harris, M. J.; and Zimmerman, B. A. 1984, "Atomic Data and Nuclear Data Tables," in preparation.

Davis, R., Jr., Cleveland, B. T., and Rowley, J. K. 1983, *Science Underground*, ed. M. M. Nieto et al. (New York: American Institute of Physics).

Dyer, P., and Barnes, C. A. 1974, *Nucl. Phys.*, **A233**, 495.

Filippone, B. W., Elwyn, A. J., and Davids, C. N. 1983, *Phys. Rev. Letters*, **50**, 412.

Fowler, W. A., Caughlan, G. R., and Zimmerman, B. A. 1975, *Ann. Rev. Astr. Ap.*, **13**, 69.

Harris, M. J.; Fowler, W. A.; Caughlan, G. R.; and Zimmerman, B. A. 1983, *Ann. Rev. Astr. Ap.*, **21**, 165.

Kettner, K. U.; Becker, H. W.; Buckmann, L.; Görres, J.; Kräwinkel, H.; Rolfs, C.; Schmalbrock, P.; Trautvetter, H. P.; and Vlieks, A. 1982, *Zs. Phys.*, **A308**, 73.

Kräwinkel, H.; Becker, H. W.; Buchmann, L.; Görres, J.; Kettner, K. U.; Kieser, W. E.; Santo, R.; Schmalbrock, P.; Trautvetter, H. P.; Vlieks, A.;

Rolfs, C.; Hammer, J. W.; Azuma, R. E.; and Rodney, W. S. 1982, *Zs. Phys.*, **A304,** 307.

Langanke, K., and Koonin, S. E. 1983, private communication.

Nagatani, K., Dwarakanath, M. R., and Ashery, D. 1969, *Nucl. Phys.*, **A128,** 325.

Nomoto, K. 1982, *Ap. J.*, **253,** 798.

Osborne, J. L.; Barnes, C. A.; Kavanagh, R. W.; Kremer, R. M.; Mathews, G. J.; and Zyskind, J. L. 1982, *Phys. Rev. Letters*, **48,** 1664.

Parker, P. D., and Kavanagh, R. W. 1963, *Phys. Rev.*, **131,** 2578.

Robertson, R. G. H.; Dyer, P.; Bowles, T. J.; Brown, R. E.; Jarmie, N.; Maggiore, C. J.; and Austin, S. M. 1983, *Phys. Rev.*, **C27,** 11.

Skelton, R. T., and Kavanagh, R. W. 1984, *Nucl. Phys.*, **A414,** 141.

Thielemann, F.-K. 1983, private communication.

Trautvetter, H. P.; Becker, H. W.; Buchmann, L.; Görres, J.; Kettner, K. U.; Rolfs, C.; Schmalbrock, P.; and Vlieks, A. E. 1983, *Verhandlunger der Deutschen Physicalischen Gesellschaft*, **VI,** 1141.

3

Status of Helium Burning of ^{12}C

C. Rolfs

Helium burning of ^{12}C via the reaction ^{12}C$(\alpha,\gamma)^{16}$O proceeds at a stellar energy near $E_0 = 0.3$ MeV. The level structure of ^{16}O reveals that there is no compound state in ^{16}O near E_0 to act as a thermal resonance in this reaction (figure 3.1). The reaction must proceed therefore via the tails of nearby states. Among the reported resonance states above the alpha-particle threshold (figure 3.1), only the $J^\pi = 1^-$, $E_x = 9580$ keV state ($E_R = 2418$ keV) has a sufficient width and capture strength to contribute to the stellar reaction rate near E_0 via the low energy tail of this E1 capture amplitude. The $E_x(J^\pi) = 7117(1^-)$ and $6917(2^+)$ keV states, which are located -45 and -245 keV below the alpha-particle threshold, respectively, decay to the ground state and can—as subthreshold resonances—contribute with their high energy tails to the stellar rate near E_0 (figure 3.1). In addition to the capture into the tails of resonances, direct capture into the ground state of ^{16}O must also be considered. In this case the capture process is an electric quadrupole (E2) transition from an initial d-partial wave into an s-orbital state. In summary, one has to consider most likely two E1 amplitudes and two E2 amplitudes for the ground state capture transition (figure 3.1).

The available information on the nuclear structure of these low-lying states in ^{16}O reveals that both configuration mixing and isospin mixing are involved in the description of these states (Ajzenberg-Selove 1982). As a consequence, pure nuclear model calculations of the ^{12}C$(\alpha,\gamma)^{16}$O reaction rate are very difficult.

For the 2.42 MeV resonance (figure 3.1), the relevant nuclear parameters extracted from various experiments are given in table 3.1. With these parameters and a knowledge of the energy dependence of the partial and thus total widths, the energy dependence of the E1 capture cross section into the ^{16}O ground state can be calculated using the Breit-Wigner formula. The results in

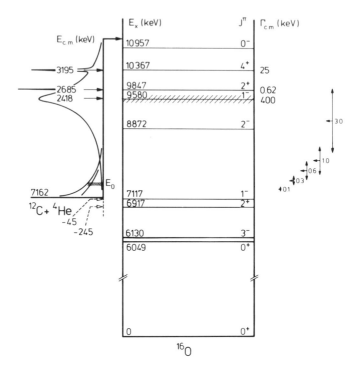

Fig. 3.1. Level scheme of ^{16}O near and above the alpha-particle threshold (Ajzenberg-Selove 1982; Kettner et al. 1982). Also shown are the energy regions of astrophysical interest for different stellar temperatures T_9 (in units of 10^9 K). In the most effective stellar energy region near $E_0 = 0.3$ MeV, the ^{12}C$(\alpha,\gamma)^{16}$O reaction rate appears to be influenced by the low energy tail of the 2.42 MeV resonance as well as the high energy tails of the two subthreshold resonances at $E_R = -45$ and -245 keV. In addition, E2 direct capture into the ground state might play an important role.

the form of the $S(E)$-factor are plotted in figure 3.2a as a dotted curve. In a like manner, one can calculate the energy dependence of the high energy tails of the two subthreshold resonances, again using the available nuclear parameters listed in table 3.1. The results are also plotted as dotted curves in figures 3.2a and 3.2b. The $S(E)$-factor for the E2 direct capture is essentially energy independent, with an absolute value proportional to the reduced alpha-particle width of the ^{16}O ground state. The dotted curve in figure 3.2b has been obtained for $\Theta_\alpha^2(\text{g.s.}) = 0.19$ (table 3.1). Since the E1 and E2 capture mechanisms both have two sources, the total cross section for E1 and E2

Table 3.1

Level Parameters of ^{16}O States Involved in the ^{12}C$(\alpha,\gamma_0)^{16}$O Stellar Reaction

	Level E_x (keV)			
	0	6917	7117	9580
J^π	0^+	2^+	1^-	1^-
E_R (keV)	. . .	-245 ± 1	-45 ± 1	2418 ± 12^h
Γ_{γ_0} (meV)	. . .	100 ± 6	57 ± 5	23 ± 3^i
	Reduced alpha-particle width Θ_α^2			
α-elastic	. . .	. . .	. . .	0.55 ± 0.02^j
α-transfer^a	0.19	1.0	0.40^f	$\equiv0.61$
(α,γ_0) data^b	0.04 ± 0.01^d	. . .	0.10^g	. . .
(α,γ_0) data^c	0.25 (+0.25,–0.08)	1.0 ± 0.2^e	0.19 (+0.14,–0.10)	0.67 ± 0.03^k

Note: Values deduced from the compilation Ajzenberg-Selove (1982) except where noted. E_R = resonance energy. Γ_{γ_0} = gamma width.

[a]Becchetti et al. (1980), relative to Θ_α^2 (9.58 MeV) = 0.61.

[b]Fit to the E1 capture data (Dyer and Barnes 1974). See also Tombrello et al. (1982).

[c]Fit to the E1 + E2 capture data (Kettner et al. 1982).

[d]See also Tombrello et al. (1982).

[e]Includes data obtained for cascade γ-ray transitions (Kettner et al. 1982).

[f]Other sources lead to values of Θ_α^2 = 0.03 to 0.40 (Kettner et al. 1982).

[g]Relative to Θ_α^2 (9.58 MeV) = 0.55 (see also Tombrello et al. 1982).

[h]From elastic scattering data (Kettner et al. 1982). See also Koonin et al. (1974). The older elastic scattering data (Jones et al. 1962) lead to 2404 ± 15 keV.

[i]The analysis of the data shown in figure 3.6b leads to 25 ± 4 meV (Kettner et al. 1982).

[j]Deduced from the total width of $\Gamma_{c.m.} = 400\pm10$ keV (Ajzenberg-Selove 1982). The recent elastic scattering data (Kettner et al. 1982) lead to $\Theta_\alpha^2 = 0.67\pm0.03$ and hence to an average value of $\Theta_\alpha^2 = 0.61$. Note that the older elastic scattering data (Jones et al. 1962) result in $\Gamma_{c.m.} = 594$ keV and hence $\Theta_\alpha^2 = 0.82$.

[k]Deduced from elastic scattering data obtained concurrently (Kettner et al. 1982).

capture will be influenced by interference effects between the two sources. These effects will be maximum at those energies where the two amplitudes are comparable, which, according to figures 3.2*a* and 3.2*b*, is in the neighborhood of $E_{c.m.}$ = 1.4 and 3.0 MeV, respectively. Of course, the interference may be either constructive or destructive, leading to either a doubling or a vanishing of the effective cross sections in these energy regions (*solid* and *dashed curves* in figures 3.2*a* and 3.2*b*). Since there are no interference effects between the El and E2 processes in the total capture cross section, the total capture cross section is simply the incoherent sum of the El and E2 yields, leading to four differing curves (figure 3.2*c*) depending on the sign of the interference effects for both the El and E2 contributions.

For the nuclear parameters of table 3.1 these calculations would indicate an *S*-factor of $S(E_0 = 0.3 \text{ MeV}) \approx$ 0.2–0.4 MeV-b, which translates into a value for the capture cross section of $\sigma(E_0 = 0.3 \text{ MeV}) \approx (3–6) \times 10^{-17}$ b. A direct measurement of this cross section is clearly beyond present technical capability. In the calculations of the curves shown in figure 3.2 one of the important parameters is the reduced alpha-particle width Θ_α^2. The values chosen were obtained from α-transfer reactions (table 3.1) and are in general in fair agreement with the corresponding values deduced from resonance data. The analysis of α-transfer reactions is very complex (Bethge 1970; Becchetti et al. 1980), and again one cannot rule out the possibility of systematic errors limiting the reliability of the resulting Θ_α^2 values (Barnes 1971; Tombrello, Koonin, and Flanders 1982).

Clearly, the best information would be obtained from a direct measurement of the cross sections for the $^{12}C(\alpha,\gamma)^{16}O$ reaction over a wide range of energies and to as low an energy as technically possible. It is clear that pushing the measurements to the lowest energies feasible, while maintaining high precision, permits a selection of the four possible solutions (figure 3.2*c*) and consequently a more accurate extrapolation to stellar energies.

Even though there are no interference effects between the El and E2 parts for the total cross sections, the differential cross sections should show interference effects because of the differing parities involved in the two multipole radiations. These interference effects show up as an asymmetry in the γ-ray angular distributions around Θ_γ = 90° (figure 3.3). While the angular distribution patterns for El and E2 capture separately do not change with beam energy (figures 3.3*a* and 3.3*b*), the pattern arising from the El and E2

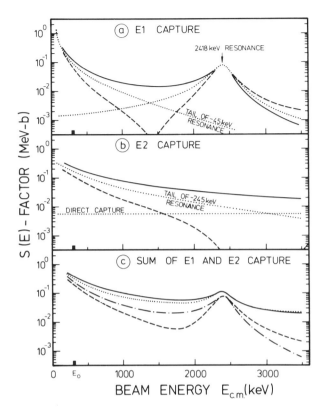

Fig. 3.2. The capture reaction $^{12}\text{C}(\alpha,\gamma_0)^{16}\text{O}$ can have contributions from both E1 and E2 amplitudes. In the case of E1 capture, the $J^\pi = 1^-$ resonances at $E_R =$ 2418 and -45 keV can both contribute to the yield (*dotted curves* in figure 3.2a), where constructive or destructive interference between both E1 sources should occur (*solid* or *dashed curves* in figure 3.2a). Similarly, the two E2 sources arising from the $E_R = -245$ keV subthreshold resonance and the direct capture process (*dotted curves* in figure 3.2b) lead to interference effects in the energy dependence of the E2 $S(E)$-factor (*solid* and *dashed curves* in figure 3.2b). The total capture cross section is simply the incoherent sum of the E1 and E2 yields leading to four differing curves (figure 3.2c) depending on the sign of the interference effects for both the E1 and E2 contributions.

interferences (figure 3.3d) will vary with beam energy because of the differing energy dependences of the individual total capture cross sections $\sigma_{\text{E1}}(E)$ and $\sigma_{\text{E2}}(E)$ (figure 3.2). Therefore, the analysis of these patterns provides the ratio of E1 and E2 total capture cross sections as a function of energy and clearly

GAMMA ANGULAR DISTRIBUTIONS FOR $^{12}C(\alpha,\gamma)^{16}O$

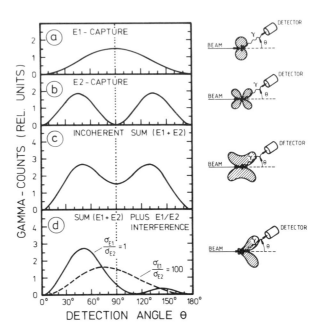

Fig. 3.3. Gamma-ray angular distributions for the $^{12}C(\alpha,\gamma_0)^{16}O$ reaction. The E1 and E2 patterns are not influenced by the interference effects of the two sources contributing to the yield for both E1 and E2 capture (figure 3.2). The incoherent sum of both patterns, shown in figure 3.3c, is not a realistic situation, since both multipole radiations interfere with each other leading to patterns asymmetric around $\Theta_\gamma = 90°$. The pattern shown in figure 3.3d assumes equal cross sections for both multipole radiations. It should be pointed out that even if both yields differ for example by two orders of magnitude, the presence of the smaller component is largely amplified through the interference term (*dashed line* in figure 3.3d). Note also that γ-ray yield measurements carried out at $\Theta_\gamma = 90°$ are only sensitive to E1 capture and this yield is also free of interference effects with the E2 capture process.

improves the precision with which one can extrapolate the data to stellar energies.

It appears from these discussions that the experimental studies of $^{12}C(\alpha,\gamma_0)^{16}O$ should consist of the measurement of total cross sections and γ-ray angular distributions over a wide range of beam energies and to as low an energy as feasible.

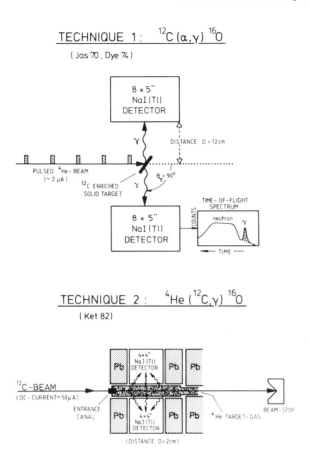

Fig. 3.4. Schematic representation of the two experimental techniques which to date have been used in measurements of the ^{12}C$(\alpha,\gamma)^{16}$O reaction.

Since the capture cross section at the peak of the 2.42 MeV resonance is only 40–50 nb, observation of interference effects in the tails of this resonance requires the measurement of cross sections considerably smaller than 1 nb. The formidable problems encountered in these measurements arise from the combination of a low γ-ray capture yield and a high neutron-induced γ-ray background. The ^{13}C$(\alpha,n)^{16}$O reaction is a prolific source of neutrons, and any ^{13}C contained in the target is therefore undesirable. As a consequence, measurements were carried out using targets of high ^{12}C isotopic purity (Jaszczak, Gibbons, and Macklin 1970; Dyer and Barnes 1974). In addition, the difference between the neutron and gamma time-of-flight was used ($=$ technique 1 in figure 3.4) to separate the γ-rays produced by neutrons and the prompt capture γ-rays (Jaszczak, Gibbons, and Macklin 1970; Dyer and Barnes 1974). In another experiment (Kettner et al. 1982), the neutron-induced γ-ray background from the $^{13}C(\alpha,n)^{16}O$ reaction was eliminated by interchanging the role of projectiles and target nuclei ($=$ technique 2 in figure 3.4), by having the ^4He target nuclei contained in a windowless gas target system and bombarding them with an intense ^{12}C beam. The results of all these measurements are shown in figure 3.5. In spite of the enormous experimental difficulties, the overall agreement of these measurements appears to be satisfactory. However, a closer inspection does reveal some differences among the data sets which should be pointed out, because even small discrepancies in the measured cross sections can lead to major differences in the extrapolations.

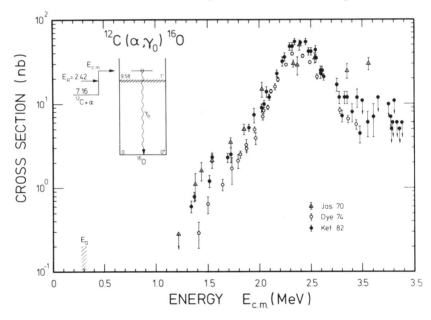

Fig. 3.5. Cross-section data for $^{12}C(\alpha,\gamma_0)^{16}O$ as reported by several investigators. Two of the data sets (Jaszczak, Gibbons, and Macklin 1970; Dyer and Barnes 1974) have been obtained by technique 1 of figure 3.4, while the more recent data (Kettner et al. 1982) involved technique 2 of figure 3.4. Note also that one data set (Dyer and Barnes 1974) represents only E1 capture.

The data of Jaszczak, Gibbons, and Macklin (1970) and Kettner et al. (1982) are in fair agreement at the lower energies. The results of Dyer and Barnes (1974) also agree very well with the other measurements in the energy dependence of the cross section for all reported data points at $E_{\text{c.m.}} \geq 1.9$ MeV. However, the absolute cross sections are about 30% smaller and the cross section values reported at $E_{\text{c.m.}} = 1.41$–1.87 MeV (Dyer and Barnes 1974) drop faster with decreasing beam energy. In the work of Dyer and Barnes (1974) only the E1 capture cross sections are reported. It is possible therefore that this discrepancy is based on the difference between partial (E1) and total (E1 + E2) cross sections, since in the work of Kettner et al. (1982) the detectors were placed in close geometry to the target (figure 3.4) resulting in angle-integrated capture yields, that is, total cross sections.

The E1 capture data of Dyer and Barnes (1974) are plotted in figure 3.6a in the form of the $S(E)$-factor. The dashed curve in this figure represents the S-factor curve resulting from the 2.42 MeV resonance alone. The data clearly show an enhancement at the lower energies, indicating constructive interference with the -45 keV subthreshold resonance (figures 3.1 and 3.2a). The data have been analyzed using various models (table 3.2) with the result that $S_{\text{E1}}(E_0 = 0.3 \text{ MeV}) \approx 0.1$–$0.2$ MeV-b for the E1 part of the capture mechanism only.

Data from the more recent investigation (Kettner et al. 1982), for the total $S(E)$-factor, are shown in figure 3.6b. Using the Breit-Wigner formalism and incorporating data from elastic scattering which was obtained concurrently (Kettner et al. 1982), the cross sections resulting from the E1 transitions were calculated and are shown as a dotted line in figure 3.6b. It is clear from the figure that the E1 capture mechanisms alone cannot account for the total capture cross sections on either side of the 2.42 MeV resonance. When the E2 capture mechanisms (figure 3.2b) are included in the analysis, the solid line in figure 3.6b is obtained, leading to an extrapolated value of $S_{\text{E1+E2}}(E_0 = 0.3 \text{ MeV}) \approx 0.4$ MeV-b. Additional information on the properties of the $J^{\pi} = 2^{+}$ subthreshold resonance obtained from cascade transitions (Kettner et al. 1982) was consistent with the above analysis. An analysis of the data of Jaszczak, Gibbons, and Macklin (1970) yielded (Barnes 1971) $S(E_0 = 0.3 \text{ MeV}) \approx 0.6$ MeV-b ($\pm 200\%$).

The extrapolated S-factor arrived at by analysis of the two sets of data (figure 3.6) differs by a factor of 2 to 4. This difference might be expected if

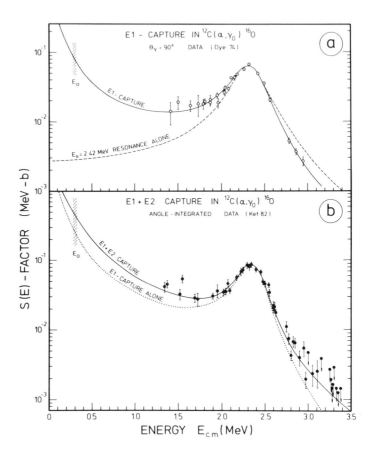

Fig. 3.6. (a) The E1 capture yield in $^{12}C(\alpha,\gamma_0)^{16}O$ as determined (Dyer and Barnes 1974) at $\Theta_\gamma = 90°$ is shown in the form of the $S(E)$-factor. The data cannot be explained by the 2.42 MeV resonance alone, but clearly require the additional contribution of the −45 keV subthreshold resonance, where the two resonances interfere constructively at energies between the two resonances. Fit parameters are given in table 3.1. (b) The observed angle-integrated γ_0-ray yields (Kettner et al. 1982) indicate the presence of E1 and E2 capture mechanisms in $^{12}C(\alpha,\gamma_0)^{16}O$ (for fit parameters, see table 3.1).

in one case the effects of the E2 capture mechanisms were not taken into account. The presence of the E2 part can also be inferred using the γ-ray angular distribution data obtained by Dyer and Barnes (1974) over a rather limited energy range by taking the ratios of the E2 and E1 contributions.

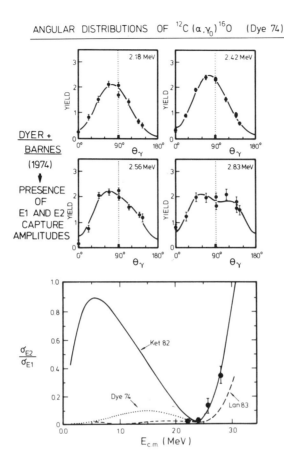

ANGULAR DISTRIBUTIONS OF ^{12}C $(\alpha,\gamma_0)^{16}$O   (Dye 74)

Fig. 3.7. Angular distributions for the ^{12}C$(\alpha,\gamma_0)^{16}$O reaction obtained (Dyer and Barnes 1974) at $E_{\text{c.m.}} = 2.18$, 2.42, 2.56, and 2.83 MeV are asymmetric around $\Theta_\gamma = 90°$ and clearly reveal the presence of both E1 and E2 capture amplitudes. The deduced ratios of E2 and E1 capture cross sections have been fit (Dyer and Barnes 1974) for the assumption of a single E2 source (*dotted line*), namely, the direct capture process (figure 3.2*b*). The four data points are also in good agreement with the σ_{E2}/σ_{E1} ratios (*solid line*) deduced from analyses of recent capture data (Kettner et al. 1982, figure 3.6*b*). Recent hybrid model calculations (Langanke and Koonin 1983) lead to ratios (*dashed line*) significantly lower than the observed values and also predict a vanishing E2 yield at low energies due to a destructive pole for the two E2 sources near $E_{\text{c.m.}} \approx 1$ MeV (figure 3.2*b*).

Table 3.2
Status of $^{12}C(\alpha,\gamma_0)^{16}O$ Rate at $E_0 = 0.3$ MeV

Data	$S(E_0 = 0.3$ MeV$)$ (MeV-b)	Analysis	Authors[*]
E1 capture	0.14 (+0.14,–0.12)	R-matrix	Dye74
(Dye74)	0.08 (+0.05,–0.04)	Hybrid R-matrix	Koo74
	0.08 (+0.14,–0.07)	K-matrix	Hum76
	$\approx$0.1–0.2	Hybrid R-matrix	Tom82
	$\approx$0.15	Hybrid R-matrix	Lan83
(E1+E2) capture	$\approx$0.42 (+0.16,–0.12)	Simple Breit-Wigner	Ket82
(Ket82)	$\approx$0.34	Hybrid R-matrix	Lan83

[*]Authors: Dye74 = Dyer and Barnes (1974); Hum76 = Humblett, Dyer, and Zimmerman (1976); Ket82 = Kettner et al. (1982); Koo74 = Koonin, Tombrello, and Fox (1974); Lan83 = Langanke and Koonin (1983); Tom82 = Tombrello, Koonin, and Flanders (1982).

These results (figure 3.7) are consistent with the analysis shown in figure 3.6*b* as well as previous analyses (Dyer and Barnes 1974) assuming only one source for E2 capture, namely, the direct capture process. Clearly, it would be desirable to have such measurements over a much wider range of energies. In spite of the enormous experimental efforts that have gone into studies of this reaction, there are still considerable uncertainties in its stellar reaction rate (table 3.2). As a consequence, neither the relative amounts of ^{12}C and ^{16}O produced by red giant stars nor the stars' subsequent evolution can be determined with great confidence. Clearly, additional experimental work is necessary to improve the knowledge of the rate of this important reaction.

References

Ajzenberg-Selove, F. 1982, *Nucl. Phys.*, **A375**, 1.

Barnes, C. A. 1971, *Advan. Nucl. Phys.*, **4**, 133.

Becchetti, F. D.; Overway, D.; Jänecke, J.; and Jacobs, W. D. 1980, *Nucl. Phys.*, **A344**, 336.

Bethge, K. 1970, *Ann. Rev. Nucl. Sci.*, **20**, 255.

Dyer, P., and Barnes, C. A. 1974, *Nucl. Phys.*, **A233,** 495. (Dye74)

Humblett, J., Dyer, P., and Zimmerman, B. A. 1976, *Nucl. Phys.*, **A271,** 210.

Jaszczak, R. S., Gibbons, J. H., and Macklin, R. L. 1970, *Phys. Rev.*, **C2,** 63 and 2452. (Jas70)

Jones, C. M.; Phillips, B. C.; Harris, R. W.; and Beckner, E. H. 1962, *Nucl. Phys.*, **37,** 1.

Kettner, K. U., et al. 1982, *Zs. Phys.*, **A208,** 73. (Ket82)

Koonin, S. E., Tombrello, T. A., and Fox, G. 1974, *Nucl. Phys.*, **A220,** 221.

Langanke, K., and Koonin, S. E. 1983, preprint. (Lan83)

Tombrello, T. A., Koonin, S. E., and Flanders, B. A. 1982, in *Essays in Nuclear Astrophysics.*, ed. C. A. Barnes, D. D. Clayton, and D. N. Schramm, p. 233 (Cambridge: Cambridge University Press).

4

Precise Stellar Neutron Capture Rates: s-Process Needs and Experimental Possibilities

F. Käppeler

Abstract

The needs for more precise neutron capture rates for a detailed understanding of s-process nucleosynthesis are reviewed. The required accuracy of $\sim 1\%$ cannot be achieved with present-day techniques, which are limited to $\sim 5\%$ by inherent systematic uncertainties. The prospects of the activation technique are discussed, and a new method for differential cross-section measurements is introduced which is based on a 4π-detector system.

4.1 Cross-Section Needs for s-Process Studies

The phenomenological approach to s-process nucleosynthesis proposed in Burbidge et al. (1957) and developed by Clayton et al. (1961) has been found to be the more successful the more accurate data for neutron capture cross sections and solar abundances became available.

At present the state of the art for neutron capture cross-section measurements gives accuracies between 5% for reasonably large cross sections and 10% for less favorable cases. Recent examples are data from Oak Ridge (Macklin 1982; Winters and Macklin 1982) and Karlsruhe (Beer, Käppeler, and Wisshak 1982; Walter 1984). A collaboration between these laboratories confirmed the estimated uncertainties by the good agreement achieved for the same isotopes (Beer and Macklin 1982). However, when cross sections are small as was the case in a recent study of ^{142}Nd (Mathews and Käppeler 1983) or when the cross sections exhibit a pronounced resonance structure as for ^{58}Fe (Käppeler, Wisshak, and Hong 1983), somewhat larger uncertainties of order 8–10% are unavoidable.

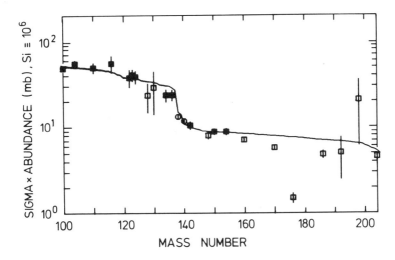

Fig. 4.1. Updated $\sigma N(A)$-curve compared to Käppeler et al. (1982). The solid line is a fit to the empirical σN-values of the pure *s*-isotopes indicated by black squares. Open symbols denote pure *s*-isotopes for which no experimental cross sections are known or which are involved in *s*-process branchings (*squares*) or isotopes with negligible *r*-process contribution (*circles*).

The situation with the abundance tables (Suess and Zeh 1973; Cameron 1973, 1982; Palme, Suess, and Zeh 1981) was difficult to judge for the nonspecialist in the field because uncertainties were not considered in these compilations. It was the merit of Anders and Ebihara (1982) to include this aspect in their detailed and comprehensive study. The quoted uncertainties for the heavy-element abundances ($Z > 20$) are typically between 5 and 10% with some exceptions above (13 cases) but also below this range (9 cases). From the excellent precision achieved in recent abundance measurements (e.g., Patchett 1983) one can expect significant improvements for the solar system abundances in general.

The good success of the phenomenological model of the *s*-process is illustrated in figure 4.1. This presents an update of the σN-curve given by Käppeler et al. (1982). It is derived with the steady flow approximation and the assumption of an exponential distribution of neutron exposures. The solid line is calculated by a least squares fit to the empirical σN-values of all *s*-only nuclei in the mass region $A > 100$ (except those belonging to *s*-process branchings and the uncertain cases ^{128,130}Xe, ^{192}Pt, and ^{198}Hg). The new cross

Table 4.1

Recent Maxwellian Averaged Capture Cross Sections
for $kT = 30$ keV of Pure s-Process Nuclei

Isotope	Maxwellian Averaged Cross Sections (mb)		Ref.[a]
	Käppeler et al. (1982)	Recent Work	
^{58}Fe	18 ± 3	14.3 ± 1.4	(1)
^{128}Xe	303 ± 151^{b}	249 ± 50^{b}	(2)
^{130}Xe	181 ± 50^{b}	153 ± 30^{b}	(2)
^{142}Nd	52 ± 10	51 ± 4	(3)
^{148}Sm	277 ± 21	277 ± 13	(4)
^{150}Sm	576 ± 190	465 ± 28	(4)

[a]References: (1) Käppeler, Wisshak, and Hong (1983), (2) Beer et al. (1983), (3) Mathews and Käppeler (1983), (4) Winters et al. (1983).
[b]Calculated cross sections.

sections used for this curve are compiled in table 4.1. The major change compared to the previous σN-curve results from the new abundances of Anders and Ebihara (1982), especially from the increased values in the rare earth region. As a consequence the σN-curve is now about 25% higher for $A > 140$, leading to a less pronounced step at the closed neutron shell $N = 82$. This in turn required a somewhat larger mean neutron exposure $(\tau_0 \sim 0.29$ mb^{-1} instead of 0.24 mb^{-1}), which means that the number of neutrons captured per ^{56}Fe seed nucleus is also larger than found previously (10 instead of 8 neutrons per ^{56}Fe seed). As the CNO abundances were also increased relative to ^{56}Fe, the above change of the neutron consumption is counterbalanced by a correspondingly larger neutron supply. In other words, the neutron balance as discussed by Almeida and Käppeler (1983) for the convective He shell burning in red giant stars is still fulfilled with the new results.

It should also be noted as an indication of the improved consistency that with the new abundances of Anders and Ebihara (1982) two heretofore discrepant σN-values—^{142}Nd and ^{154}Gd—now fit the calculated σN-curve very well.

However, in spite of the overall agreement between the calculated σN-curve and the empirical values of s-only isotopes, there are many problems

Table 4.2

Experimental Values for the 30 keV Maxwellian Averaged
Cross Sections for ^{148,149,150}Sm

Isotope	Maxwellian Averaged Cross Section (mb)		
^{148}Sm	257 ± 50 (1)	281 ± 23 (2)	277 ± 12 (3)
^{149}Sm	1620 ± 280 (1)	2490 ± 200 (4)	1489 ± 65 (3)
^{150}Sm	370 ± 72 (1)	690 ± 51 (2)	465 ± 28 (3)

References: (1) Macklin, Gibbons, and Inada (1963),
(2) Kononov et al. (1978), (3) Winters et al. (1983),
(4) Mizumoto (1981).

which require more precise data. One such example is the question of the
validity of the "local approximation," i.e., whether the σN-values of neighboring isotopes are really equal. This feature can best be tested for those cases
where two or more s-only isotopes of the same element occur. The existing
data for the examples of ^{122,123,124}Te and ^{134,136}Ba confirm the local approximation to $\sim 10\%$ corresponding to the related cross-section uncertainties.
The third example, ^{148,150}Sm, was confused by an 80% discrepancy for the
^{150}Sm cross section. For this reason, Winters et al. (1983) performed a new
and accurate measurement of the capture cross sections of ^{148,149,150}Sm in the
neutron energy range from 4 to 250 keV. Their Maxwellian averages for
$kT = 30$ keV are compared in table 4.2 with previous values. It is remarkable that the two recent measurements on ^{149}Sm (Mizumoto 1981) and ^{150}Sm
(Kononov et al. 1978) show large discrepancies with respect to the new data
which are quoted with a precision of 4.5%, whereas the early measurements
by Macklin, Gibbons, and Inada (1963) agree within their uncertainties.

The new data of Winters et al. (1983) yield a ratio

$$\frac{\sigma N(^{148}\text{Sm})}{\sigma N(^{150}\text{Sm})} = 0.91 \pm 0.03 \ ,$$

which is significantly smaller than unity. Among the various possible reasons
for this result they consider a branching of the s-process capture path at
$A = 147,148$ as the most likely possibility. By this branching ^{148}Sm is partly
bypassed by the s-process flow as is illustrated in figure 4.2. The beta decay
rates of the involved unstable isotopes (^{147}Nd, ^{147,148}Pm) seem to be almost
independent of the supposed s-process temperature so that this branching can

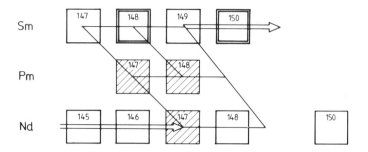

Fig. 4.2. The *s*-process branching at $A = 147,148$.

be analyzed to yield an estimate for the *s*-process neutron density. Winters et al. (1983) find in the steady flow approximation

$$n_n = (1.3 \pm 0.5) \times 10^8 \text{ cm}^{-3} \ .$$

The associated uncertainty is due mostly to the uncertainty in the cross-section ratio. Hence, more precise cross sections are needed to improve this situation.

Besides this example, there are many more *s*-process problems for which better data are needed, the most obvious being again the confirmation of the local approximation at the remaining cases of Te and Ba. In table 4.3 these problems are summarized and briefly commented on. Section 4.2 deals with the present experimental methods and their limitations. For the activation technique some possible improvements are discussed. Finally, in § 4.3 a novel technique for capture cross-section measurements is proposed and outlined in some detail.

4.2 Present Techniques

4.2.1 Differential Methods

In figure 4.3 the signature of a capture event is sketched schematically. After neutron capture the highly excited compound system returns to the ground state by emission of a gamma-ray cascade which is determined only by statistics. (The compound nucleus mechanism used here dominates the other capture mechanisms such as valence or direct capture for almost all isotopes on the *s*-process path.) The very many possible gamma transitions in the

Table 4.3
Various Aspects of *s*-Process Nucleosynthesis Where
Precise Capture Cross Sections Are Required

Aspect	Comment
Local approximation	Probably ^{122,123,124}Te are most promising because ^{134,136}Ba might also be disturbed by a branching, like ^{148,150}Sm. Note that thermal effects on ^{123}Te might be found through their possible influence on the ^{123}Te/^{123}Sb ratio
σN-curve	i) Better data improve the fit to empirical σN-values and hence the quality of the σN-curve ii) *s*-process details may emerge from significant deviations of empirical σN-values from the model curve iii) *s*-process abundances $N_s = (\sigma N)^{calc}/\sigma$ depend directly on accurate cross sections. This is important for determining *r*-process residuals and for analysis of isotopic anomalies
s-process branchings	*s*-process branchings can be identified by comparison of the respective empirical σN-values and the σN-curve. Only when a significant difference shows up can a reliable branching ratio be derived
Pulsed *s*-process	Abundance characteristics due to a pulsed *s*-process (with time-dependent neutron density and temperature) can be quantified only with precise cross sections because these effects are probably small
Chronometers	Although it seems that there are nontrivial problems with practically all radioactive cosmic clocks (e.g., Yokoi, Takahashi, and Arnould 1983), better cross sections would help to eliminate at least that uncertainty from the problem. This holds for the three cases of ^{176}Lu, ^{87}Rb, and ^{187}Os.

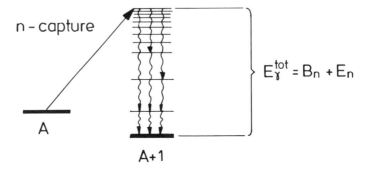

Fig. 4.3. Characteristics of a neutron capture event.

decay of the compound nucleus lead to a smooth capture gamma-ray spectrum without prominent, isolated peaks which could be used for identification of capture events by high resolution spectrometry, e.g., with Ge(Li)-detectors. The only well defined quantity in this reaction is the total gamma-ray energy of the cascade

$$E_\gamma^{tot} = B_n + E_n \ ,$$

where B_n is the neutron separation energy. The kinetic energy of the incoming neutron E_n is almost negligible in the energy range of interest to the s-process, which peaks at $E_n = 30$ keV. Consequently, the best measurement technique would be to detect all gamma-rays in the cascade to obtain a signal proportional to E_γ^{tot}. An ideal detector covering the entire solid angle of 4π with 100% efficiency and good energy resolution would then yield a spectrum of capture events as indicated in figure 4.4a. A first approximation to this technique was achieved by means of large liquid scintillator tanks of 300–3000 l volume. The efficiency of these detectors was of order 70%, but they suffered from several drawbacks: (i) their large volume led to a significant background from cosmic rays, and, more severe, (ii) their energy resolution was poor, and (iii) they were sensitive to neutrons scattered in the sample. These neutrons were thermalized in the scintillator and subsequently captured in hydrogen. Because neutron scattering is approximately an order of magnitude more likely than neutron capture, the 2.2 MeV gamma-ray line from capture in hydrogen was overwhelming the spectra taken with scintillator tanks. A typical example for such a case is shown in figure 4.4b. In practice an electronic threshold had to be used in order to discriminate events

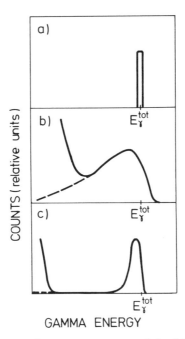

Fig. 4.4. (*a*) Spectrum of capture events recorded with an ideal 4π detector. (*b*) Typical spectrum taken with a liquid scintillator tank. The peak around E_γ^{tot} is broadened and obscured by an intense background. (*c*) Expected spectrum for a 4π detector of BaF$_2$.

with small pulse heights. The corresponding correction for the undetected capture events below this threshold had to be determined from an extrapolation of the spectrum to zero pulse height. This procedure was affected by large systematic uncertainties of 5–8%, thus limiting the overall accuracy of this method to $\sim$10% (Chrien 1975).

In order to circumvent the background problems with scintillator tanks an alternative detector was developed by Moxon and Rae (1963). This very simple setup is shown in figure 4.5*a*. It consists of a graphite disk acting as a converter of gamma-rays into electrons, followed by a thin plastic scintillator and a photomultiplier. The idea was to obtain an efficiency proportional to gamma-ray energy: electrons from low energy gamma-rays can reach the scintillator only from a thin layer of the converter whereas high energy gamma-rays produce electrons which are energetic enough to penetrate the entire converter. As the efficiency of this Moxon-Rae detector is small ($\lesssim$ 1%), at most

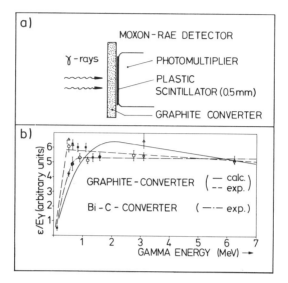

Fig. 4.5. (a) Schematics of a Moxon-Rae detector. (b) Efficiency of a Moxon-Rae detector versus gamma-ray energy (Wisshak and Käppeler 1981).

one gamma-ray per cascade is detected and the efficiency is in good approximation (Chrien 1975):

$$\epsilon_{\text{casc}} = \sum_{i=1}^{\nu} \epsilon_i\,(E_{\gamma i}) \quad ; \quad \epsilon_i\,(E_{\gamma i}) = kE_{\gamma i}$$

$$= k \sum_{i=1}^{\nu} E_{\gamma i} = kE_{\gamma}^{\text{tot}} \quad .$$

It is independent of the capture gamma-ray cascade and only determined by the total energy E_{γ}^{tot}. With the setup of figure 4.5a the proportionality between gamma-ray energy and efficiency is almost fulfilled as is shown in figure 4.5b except for very low energies where the efficiency drops very steeply. This can be a significant advantage for the investigation of radioactive targets because low energy gamma-rays from the natural decay are automatically suppressed. This has been demonstrated at several examples by Wisshak and Käppeler (1983 and references therein). These authors also studied the systematic uncertainties of Moxon-Rae measurements in great detail and found that accuracies of typically 5% can be achieved (Reffo et al. 1983).

A generalization of the Moxon-Rae detector was developed by Macklin and Gibbons (1967). In this approach the converter of figure 4.5a is eliminated and the scintillator is increased to $\sim$10 cm thickness. The volume of 1 l is small enough that one can afford to replace the usual scintillator by a deuterated scintillator to avoid the problem of hydrogen capture. This so-called total energy detector offers a much better efficiency per cascade of $\sim$20% but no longer has an efficiency proportional to the gamma-ray energy. As was shown experimentally the gamma-ray energy can be approximated by the pulse height of this detector and a weighting function $W(E_\gamma)$ can be applied to each event in order to restore the relation $\epsilon \sim E_\gamma$ artificially. This type of detector was extensively used in capture measurements, mostly in Oak Ridge by Macklin and collaborators (see Winters and Macklin 1982 as a recent example) but also in many other laboratories (e.g., in Karlsruhe, Livermore, Geel, and Tokyo). Similar to the Moxon-Rae detector one finds that inherent systematic uncertainties restrict the accuracy of this technique to $\sim$5%. This value seems to be a realistic limit for the differential techniques available at present which have been optimized over more than a decade. Therefore a further improvement of the neutron capture rates for the s-process can be achieved only by new techniques. A proposal for such a detector (Käppeler, Schatz, and Wisshak 1983) is outlined in § 4.3.

4.2.2 Activation Technique

The merits of the activation technique in conjunction with the ^{7}Li(p,n)-reaction for the determination of stellar capture rates were outlined by Beer and Käppeler (1980). The main point is the possibility of producing a neutron energy spectrum very similar to that during the s-process. This advantage is added to the well-known features of the activation technique (i.e., its selectivity, sensitivity, and accuracy), resulting in a very successful method for neutron capture cross-section measurements.

The setup during neutron irradiation is shown on the left side of figure 4.6. The energy of the incoming protons is chosen slightly above the threshold of the ^{7}Li(p,n)-reaction. Then all neutrons are collimated by the reaction kinematics in a forward cone of 120° opening angle. The angle integrated neutron spectrum imitates almost perfectly a Maxwellian energy spectrum for a thermal energy of $kT = 25$ keV as is illustrated on the right side of figure 4.6. The sample is located very close to the neutron target and is sandwiched between two gold foils which are used as a cross-section standard. A neutron monitor at a distance of 60 cm from the neutron target records the time

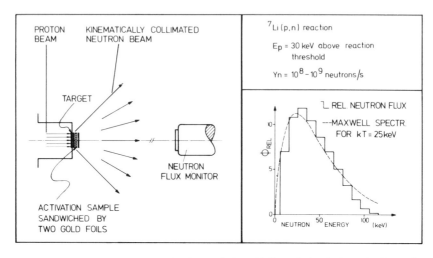

Fig. 4.6. Principle of the activation technique (*left*), and simulation of a Maxwellian energy spectrum by the ^{7}Li(p,n) reaction (*right*).

dependence of the neutron yield as the target degrades during the activation. From this time dependence, the fraction of activated nuclei which decay during irradiation is calculated. The resulting correction can be significant if the half-lives of the investigated isotope and the gold reference are very different.

After irradiation the activated samples are counted with high resolution gamma or beta detectors as a function of time. This allows a check for proper background subtraction in the activity measurement because only then the correct half-lives can be reproduced. The features of the activation technique may be illustrated at the example given in figure 4.7. It shows the gamma-ray spectrum from a ^{48}Ca sample after irradiation by neutrons from the ^{7}Li(p,n) reaction (Käppeler, Walter, and Mathews 1983). The sample consisted of a 43 mg CaCo$_3$ tablet (6 mm diameter, 82% ^{48}Ca) between two gold foils (15 mg each). The irradiation and counting time of 1000 s corresponds roughly to two half-lives of ^{49}Ca. The gamma-ray spectrum shows clearly the 411 keV line from ^{198}Au, the background line from ^{40}K, and the 3089 keV line from ^{49}Ca. Although statistics are rather poor after one cycle, it is obvious that backgrounds are almost negligible. This means that by repeated irradiations an accurate cross-section ratio to ^{197}Au can be obtained. The good energy resolution in the gamma-ray spectrum allows identification of the gamma-ray transition in the particular isotope of interest which makes the

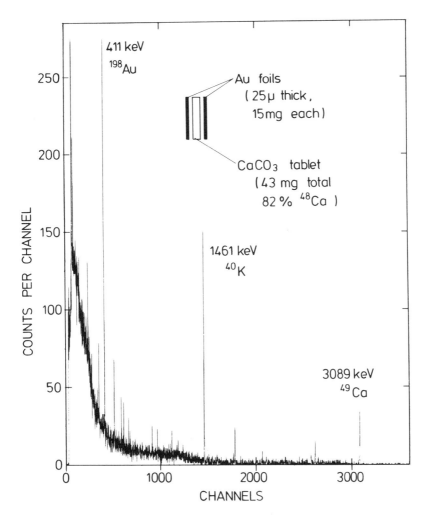

Fig. 4.7. An example of the sensitivity of the activation technique: gamma-ray spectrum obtained for the difficult case of ^{48}Ca. The inset shows the activated sample.

method selective: no highly enriched samples are required. The method is also very sensitive because of the low background: rather small sample amounts are sufficient, and very small cross sections can be determined. For example, the ^{48}Ca cross section was found to be ~ 2 mb only. And finally, the method is rather accurate as is summarized in table 4.4. The spectroscopic

Table 4.4
Systematic Uncertainties of the
Activation Technique

Effect	Uncertainties (%)	
	Present	Future
Efficiency of activity counting	2–5	1
Self-absorption	<1	<0.5
Sample mass	1–10	<1
Decay during activation	<3	<1
Relative intensity of detected radiation	1–10	?
Gold standard cross section	2.5–10	?

information for the decay of the activated isotope which has to be taken from the literature often limits the accuracy, but in principle activation measurements could reach a precision of $\sim$2%. Hence, this method is indispensable for s-process investigations, especially because of its high sensitivity.

4.3 A New 4π Detector

4.3.1 Principle

For a significant improvement of the differential techniques described in § 4.2.1 one has to return to the idea of detecting the total gamma-ray energy of the capture cascade (Käppeler, Schatz, and Wisshak 1983). Figure 4.4a shows the gamma-ray spectrum for the ideal detector with 100% efficiency. While such a detector is only roughly approximated by the liquid scintillator tank (figure 4.4b), the use of scintillation crystals appears more promising because of the much better energy resolution. The gamma-ray spectrum expected for a 4π detector of scintillation crystals is sketched in figure 4.4c. Most of the events are recorded with their full or almost full gamma-ray energy. Only a rather small fraction appears at smaller energies, because part of the gamma-rays escape from the detector or are absorbed in detector cannings or in the sample.

Table 4.5
Properties of Scintillation Crystals for Gamma-Rays

Crystal	Energy Resolution (%) (^{137}Cs)	Time Resolution (ns) (^{60}Co)	Density (g cm^{-3})	Neutron Sensitivity
NaI	8	3	3.7	–
BGO	16	~5	7.1	+
BaF$_2$	10–11	$\lesssim$0.5	4.9	+

4.3.2 Crystals

The possible scintillation crystals and their relevant properties are listed in table 4.5. One finds that the most common material, NaI, could not be used in a detector for neutron measurements in spite of the good values for energy and time resolution. The problem with NaI is the very high neutron sensitivity due to the large capture cross section of iodine. Furthermore, NaI has the lowest density so that rather thick crystals would be needed to obtain an efficiency close to 100% which would further increase the neutron sensitivity. The other two possibilities of table 4.5, bismuth germanate, $Bi_4Ge_3O_{12}$, or BGO for short, and barium fluoride, BaF$_2$, exhibit almost equal but much smaller neutron sensitivities. While BGO has the highest density, its energy and time resolution are rather poor. Therefore, at present, the best material for a neutron capture detector is BaF$_2$. The properties of this material were for a long time considered to be inferior to NaI until recently, when a very attractive feature came to light: the scintillation light of BaF$_2$ contains an extremely fast component (rise time = 100 ps, FWHM = 550 ps) which is emitted in the ultraviolet at 220 nm (Laval et al. 1983). This fast component can be detected only if photomultiplier tubes with high quality quartz windows are used and was therefore overlooked in previous investigations. It allows for excellent timing while the energy resolution of BaF$_2$ is close to that of NaI. As the costs are also much lower than for BGO, barium fluoride is the material of choice for the Karlsruhe 4π detector.

4.3.3 Configuration

Because the size of the available crystals is limited, a 4π detector must be composed of many elements. So far, two concepts for a 4π geometry have been pursued. A honeycomb structure of equally shaped hexagonal crystals

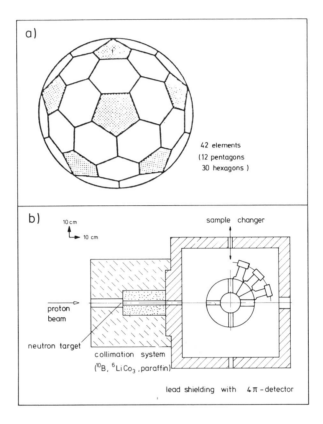

Fig. 4.8. (a) Configuration of the proposed 4π detector for capture cross-section measurements consisting of 12 pentagons and 30 hexagons. (b) Schematic setup of the 4π detector at the accelerator.

arranged in axial symmetry offers the advantage of flexibility (Twin 1983). The arrangement can be changed by inclusion of more and more elements, all of the same shape. The second approach, a crystal ball type of detector, is shown in figure 4.8a. This geometry (Habs, Stephens, and Diamond 1979) is based on elements all covering the same solid angle with respect to the sample, thus facilitating multiplicity measurements. The resulting spherical shell is composed of 12 pentagons and different discrete numbers of hexagons. For the Karlsruhe detector 42 elements were chosen, mostly because of the limited size of the available BaF_2 crystals. A sketch of the setup at the accelerator is given in figure 4.8b, indicating the 4 through holes in the detector for the

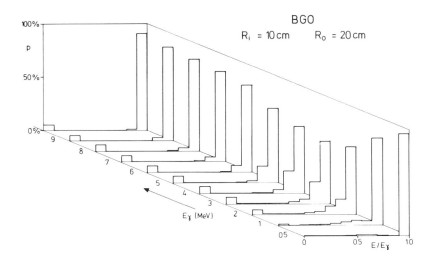

Fig. 4.9. Response of a 4π detector to monoenergetic gamma-rays.

neutron beam and the sample changer. These openings and the spacings between the single elements reduce the 4π solid angle by less than 5%.

4.3.4 Detector Properties

The calculated response of a 10 cm thick 4π detector of BGO (which is equivalent to the proposed BaF_2 detector with 17.5 cm thickness) is shown in figure 4.9 for monoenergetic gamma-rays in steps of 1 MeV (Wisshak, Käppeler, and Schatz 1983). Each spectrum is divided into 10 energy bins to simulate an energy resolution of 10%. The detection probability p is calculated by following up to three interactions of the gamma-rays. Thereafter, it was conservatively assumed that the remaining gamma-rays escape from the detector. The full energy peak contains 60–90% of the efficiency for energies between 2 and 10 MeV. In cases where the peak efficiency is low most of the remaining intensity is found in neighboring bins. The first bin contains events where the gamma-ray escapes without interaction. This probability is $\sim$6% for most energies while it is less than $\sim$1% in intermediate bins.

From this information the efficiency for neutron capture events can be derived. This was calculated for three extreme examples (^{56}Fe, ^{197}Au, and ^{241}Am) characterized by very different gamma-ray cascades with respect to total energy, E_γ^{tot}, and to multiplicity. Figure 4.10 presents the resulting

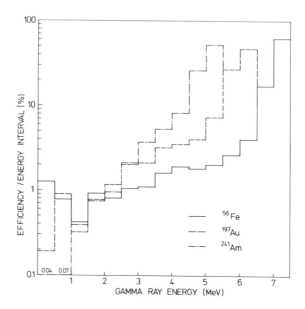

Fig. 4.10. Calculated efficiency for neutron capture cascades from ^{56}Fe, ^{197}Au, and ^{241}Am, which represent extremely different cases. Note that for a threshold of 3 MeV, all isotopes have the same efficiency of 95%.

efficiency per gamma-ray energy interval. One finds a total efficiency of 95% for all cases if a realistic threshold of 3 MeV is used for background suppression. This is remarkable as it means that the difference in efficiency even for very different capture cascades is only ∼1%, leading to a very small systematic uncertainty in the measurements.

As a last point, calculated TOF spectra of a typical Van de Graaff experiment are compared for BGO and BaF$_2$ in figure 4.11. For convenience, the TOF scale has been converted into a neutron energy scale. The much better signal to background ratio of the BaF$_2$ detector results from its superior time resolution. Although the flight path was reduced by a factor of 2 compared to BGO, the spectra taken with BaF$_2$ still show a 2 times better resolution. The dashed line indicates the background due to scattered neutrons. It occurs delayed by the additional neutron flight time between sample and crystals, which means that there is an almost background-free region close to the maximum neutron energy. Again, the better time resolution of BaF$_2$ is of

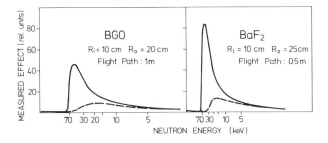

Fig. 4.11. Calculated time-of-flight spectra for a 4π detector of BaF_2 and BGO. The better time resolution of BaF_2 results in a much better signal to background ratio. The dashed line gives an upper limit for the background from scattered neutrons.

advantage for optimal discrimination. Note that the background due to scattered neutrons is equal for the two crystals; it was calculated in a straightforward way, neglecting the possibility of a further reduction by various experimental restrictions.

This section can be summarized as follows:

1. The proposed 4π detector allows for a specific registration of neutron capture events with almost 100% efficiency. This means a 10–50 times better signal to background ratio compared to present methods. The uncertainty of the absolute efficiency for capture events will be somewhat less than 1% and significantly smaller in ratio measurements.

2. These features of the new detector allow for much shorter measuring times, which in turn enable the careful experimental investigation of remaining uncertainties. These are related to the capture sample itself and concern neutron multiple scattering and self-shielding as well as gamma-ray self-attenuation.

3. From the properties of the new detector and from the experience with present techniques an overall accuracy of ~1% can be expected from this new method of neutron capture cross-section measurements.

4.4 Conclusions

Further improvements of the phenomenological *s*-process model require more precise stellar neutron capture rates. The discussion of various *s*-process problems indicates that an accuracy of $\sim$1% is needed for these data, especially for pure s-isotopes, the chronometric pairs, s-process branchings, and the interpretation of isotopic anomalies.

Among the present experimental methods, only the activation technique can be expected to yield results with the required precision whereas the differential methods seem to be limited to $\sim$5%. The activation technique is important because measurements of even small cross sections can be made with uncertainties of a few percent. The limit of this technique—often determined by the spectroscopic information involved—may be reduced to $\sim$2% in the future.

To replace the present differential methods, a new technique is proposed using a 4π gamma detector with good time and energy resolution combined with a well-defined efficiency of almost 100%. With such a detector, backgrounds and hence measuring times are drastically reduced, so that remaining uncertainties can be determined experimentally. From our design studies the new technique can be estimated to achieve the desired 1% accuracy level for stellar neutron capture rates.

References

Almeida, J., and Käppeler, F. 1983, *Ap. J.*, **265**, 417.

Anders, E., and Ebihara, M. 1982, *Geochim. Cosmochim. Acta*, **46**, 2363.

Beer, H., and Käppeler, F. 1980, *Phys. Rev.*, **C21**, 534.

Beer, H., Käppeler, F., Reffo, G., and Venturini, G. 1983, *Ap. Space Sci.*, **97**, 95.

Beer, H., Käppeler, F., and Wisshak, K. 1982, *Astr. Ap.*, **105**, 270.

Beer, H., and Macklin, R. L. 1982, *Phys. Rev.*, **C26**, 1404.

Burbidge, G. R., Burbidge, E. M., Fowler, W. A., and Hoyle, F. 1957, *Rev. Mod. Phys.*, **29**, 54.

Cameron, A. G. W. 1973, *Space Sci. Rev.*, **15**, 129.

——. 1982, in *Essays in Nuclear Astrophysics*, ed. C. A. Barnes, D. D. Clayton, and D. N. Schramm (Cambridge: Cambridge University Press).

Chrien, R. E. 1975, in *Nuclear Cross Sections and Technology*, ed. R. A. Schrack and C. D. Bowman, **vol. 1**, p. 139 (Washington, D.C.: NBS Special Publication 425).

Clayton, D. D., Fowler, W. A., Hull, T. E., and Zimmerman, B. A. 1961, *Ann. Phys.*, **12**, 331.

Habs, D., Stephens, F. S., and Diamond, R. M. 1979, Report LBL-8945.

Käppeler, F., Beer, H., Wisshak, K., Clayton, D. D., Macklin, R. L., and Ward, R. A. 1982, *Ap. J.*, **257**, 821.

Käppeler, F., Schatz, G., and Wisshak, K. 1983, Report KfK-3472.

Käppeler, F., Walter, G., and Mathews, G. J. 1983, in preparation.

Käppeler, F., Wisshak, K., and Hong, L. D. 1983, *Nucl. Sci. Eng.*, **84**, 234.

Kononov, V. N., Yurlov, B. D., Poletaev, E. D., and Timokhov, V. M. 1978, *Soviet J. Nucl. Phys.*, **27**, 5.

Laval, M., Moszynski, M., Allemand, R., Cormoreche, E., Guinet, P., Odru, R., and Vacher, J. 1983, *Nucl. Instr. Meth.*, **206**, 169.

Macklin, R. L. 1982, *Nucl. Sci. Eng.*, **81**, 520.

Macklin, R. L., and Gibbons, J. H. 1967, *Phys. Rev.*, **159**, 1007.

Macklin, R. L., Gibbons, J. H., and Inada, T. 1963, *Nature,* **197**, 369.

Mathews, G. J., and Käppeler, F. 1983, *Ap. J.* in press.

Mizumoto, M. 1981, *Nucl. Phys.*, **A357**, 90.

Moxon, M. C., and Rae, E. R. 1963, *Nucl. Instr. Meth.*, **24**, 445.

Palme, H., Suess, H. E., and Zeh, H. D. 1981, in *Landolt-Börnstein*, n.s. VI/2, chaps. 3, 4.

Patchett, J. 1983, paper read at 46th Annual Meeting of the Meteoritical Society, Mainz (Federal Republic of Germany), 5–9 September.

Reffo, G., Fabbri, F., Wisshak, K., and Käppeler, F. 1982, *Nucl. Sci. Eng.*, **80**, 630.

——. 1983, *Nucl. Sci. Eng.*, **83**, 401.

Suess, H. E., and Zeh, H. D. 1973, *Ap. Space Sci.*, **23**, 173.

Twin, P. J. 1983, paper read at International Conf. on Nuclear Physics, Florence (Italy), 29 August–3 September.

Walter, G. 1984, Ph.D. thesis, University of Heidelberg; Report KfK-3706.

Winters, R. R., Käppeler, F., Wisshak, K., Reffo, G., and Mengoni, A. 1983, in preparation.

Winters, R. R., and Macklin, R. L. 1982, *Phys. Rev.,* **C25**, 208.

Wisshak, K., and Käppeler, F. 1981, *Nucl. Sci. Eng.*, **77**, 58.

——. 1983, *Nucl. Sci. Eng.*, **85**, 251.

Wisshak, K., Käppeler, F., and Schatz, G. 1983, Report KfK-3580; *Nucl. Instr. Meth.*, **221**, 385.

Yokoi, K., Takahashi, K., and Arnould, M. 1983, *Astr. Ap.*, **117**, 65.

5

Primordial Nucleosynthesis: Looking Good!

Gary Steigman

5.1 Introduction

In this paper I will present an overview of the observational data from which the present (or presolar) abundances of deuterium, helium-3, helium-4, and lithium-7 are derived. I will also discuss the methods—and associated problems—for deriving the primordial abundances. Estimates are made of the primordial abundances—including their uncertainties—and these are compared with the predictions of primordial nucleosynthesis as calculated in the standard (i.e., simplest) hot big bang model. Subject to the uncertainties in the observational data and in the techniques for inferring the primordial abundances, the standard model is in excellent agreement with the current data. Consistency of the standard model with observations is obtained for the nucleon abundance ($\eta \equiv$ ratio of nucleons to photons) in the range 3–4 $\lesssim \eta_{10} \lesssim$ 7–10 ($\eta_{10} \equiv 10^{10}\eta$). In the standard model it is assumed that there are three two-component neutrinos ($N_\nu = 3$; ν_e, ν_μ, ν_τ); consistency with the abundance of ^{4}He is maintained for $N_\nu \lesssim 4$.

5.2 Predicted Primordial Abundances

In figure 5.1 and table 5.1 are shown the abundances of D, ^{3}He, ^{4}He, and ^{7}Li calculated as a function of the nucleon abundance in the standard model ($N_\nu = 3$). The abundances of D, ^{3}He, and ^{7}Li are by number with respect to hydrogen; the abundance by mass of ^{4}He is presented ($Y_p = 4y_p(1 + 4y_p)^{-1}$). The calculations (Yang et al. 1984, hereafter YTSSO) are for an assumed neutron half-life of $\tau_{1/2} = 10.6$ min. The "error" bar in figure 5.1 shows the effect on the predicted value of Y_p due to the uncertainty in $\tau_{1/2}$; for $\Delta\tau_{1/2} \approx \pm0.2$ min., $\Delta Y_p \approx \pm0.003$. The three curves for Y_p shown in figure 5.1 are for three assumed values for N_ν, the effective number of two-component light (< 1 MeV) neutrinos. In the standard model, $N_\nu = 3$. In

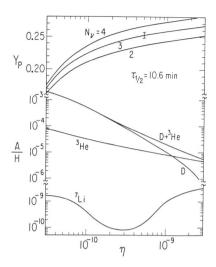

Fig. 5.1. The abundance of ^{4}He (by mass) and of D, ^{3}He, and ^{7}Li (by number relative to H) as a function of the nucleon to photon ratio (η) for three two-component neutrinos ($N_\nu = 3$) and a neutron half-life ($\tau_{1/2}$) of 10.6 min. For ^{4}He the results for $N_\nu = 2, 3, 4$ are shown and the size of the "error" bar shows the range in Y_p which corresponds to $\Delta \tau_{1/2} = \pm 0.2$ min.

the unlikely case that the τ-neutrino is heavy ($\gg 1$ MeV), $N_\nu = 2$. If there is another "flavor" of light neutrino (and/or other light "inos"), $N_\nu \geq 3$; the effect of $N_\nu = 4$ is shown. For $2 \lesssim \eta_{10} < 10$ and $N_\nu \lesssim 4$, the calculations (YTSSO) show that $\Delta Y_p \approx 0.013(N_\nu - 3)$.

Before turning to the observational data, it is worthwhile to call attention to some important trends in the predicted abundances. For ^{7}Li note that for the nucleon abundance in the range $\eta_{10} = 1\text{-}10$, the predicted abundance is in the "valley": ^{7}Li/H $\approx (0.8\text{-}9) \times 10^{-10}$. Both D and ^{3}He decrease in abundance with increasing nucleon abundance; D/H varies (with η) much more rapidly than ^{3}He/H. As η_{10} increases from 1 to 10, ^{3}He/H decreases from $\sim 3 \times 10^{-5}$ to $\sim 1 \times 10^{-5}$; over the same range, the deuterium abundance is well fit by $10^5(\text{D/H}) \approx 48\eta_{10}^{-1.6}$. The ^{4}He abundance varies very slowly with η (note the *linear* scale for Y_p in figure 5.1). For $N_\nu = 3$ and $\tau_{1/2} = 10.6$ min., the helium-4 abundance (corrected for the finite temperature effects discussed by Dicus et al. 1982) is well fit (to better than ± 0.001) by $Y_p = 0.230 + 0.011 \ln \eta_{10}$ for $2 \lesssim \eta_{10} \lesssim 10$. As discussed by YTSSO (see also

Table 5.1

Primordial Abundances of D, ^{3}He, ^{4}He, and ^{7}Li

$10^{10}\eta$	10^5(D/H)	10^5(D + ^{3}He)/H	$10^{10}(^7$Li/H)	Y_p
1	49	53	4.4	0.225
1.5	25	28	1.8	0.233
2	16	18	1.1	0.237
3	8.1	9.7	0.76	0.243
4	5.1	6.5	1.0	0.246
5	3.6	4.8	1.7	0.248
6	2.7	3.8	2.7	0.250
7	2.1	3.1	3.9	0.252
8	1.7	2.6	5.3	0.253
9	1.4	2.3	6.9	0.254
10	1.1	2.0	8.6	0.255
15	0.48	1.2	17	0.258
20	0.23	0.87	25	0.261

Note: Abundances are given for $N_\nu = 3$ and $\tau_{1/2} = 10.6$ min.

Beaudet and Reeves 1983) the calculated abundances of ^{4}He should be accurate to $\Delta Y_p \approx \pm 0.002$ and those of D and ^{3}He to $\pm 10\%$ (for fixed values of η, N_ν, and $\tau_{1/2}$). However, it is worth emphasizing that, due to existing uncertainties in the nuclear reaction cross-section data, the predicted abundance of ^{7}Li is not likely to be more reliable than a factor of 2; for ^{7}Li the accuracy of the observational data may exceed that of the predictions of the standard model.

5.3 Solar System Abundance Estimates

The Sun and the solar system provide a sample of material whose abundances may be determined from direct measurements (e.g., meteorites, lunar rocks, solar wind, etc.) and/or from indirect observational data (e.g., Jupiter, the Sun). The accumulated data may then be used to infer the presolar abundances of the light elements of interest to us. As advertised, the following discussion is an overview; see YTSSO and the references therein for the details.

5.3.1 Meteorites and the Solar Wind

The meteoritic abundance of ^{7}Li is within a factor of 2 of ^{7}Li/H = 1.2 × 10^{-9} (Mason 1979). This value is consistent with determinations of the ^{7}Li abundance in the interstellar medium (Vanden Bout et al. 1978) and in the atmospheres of Population I stars (Zappala 1972; Duncan 1981). The lower abundance in the Sun is evidence for ^{7}Li having been mixed down from the surface and burned away in the last 4.5 billion years.

The gas-rich meteorites, the lunar soil, and the foil placed on the Moon by the *Apollo* astronauts provide, presumably, a sample of solar wind–implanted ^{3}He. During the approach of the Sun to the main sequence, any presolar deuterium would have been burned to ^{3}He so that solar wind ^{3}He provides a sample of presolar ^{3}He *plus* D. In gas-rich meteorites Jeffrey and Anders (1970) found $[(D + {}^3He)/{}^4He]_{pre\odot} = (3.8 \pm 0.4) \times 10^{-4}$; Black (1972) found $[(D + {}^3He)/{}^4He]_{pre\odot} = (3.9 \pm 0.3) \times 10^{-4}$. From the solar wind–implanted lunar foil, Geiss and Reeves (1972) derived $[(D + {}^3He)/{}^4He]_{pre\odot} = (4.3 \pm 0.3) \times 10^{-4}$.

The carbonaceous chondrites provide a sample of the primitive material of the presolar nebula. The ^{3}He/^{4}He ratio is found to vary in carbonaceous chondrites; it is generally *lower* than that found for gas-rich meteorites. Anders et al. (1970) and Black (1972) identify the *lowest* ^{3}He/^{4}He ratios with material in which the deuterium has not been burned to helium-3. If this interpretation is correct, then the lowest ^{3}He/^{4}He ratios should provide good estimates of the presolar abundance of ^{3}He. For $({}^3He/{}^4He)_{pre\odot}$, Anders et al. (1970) find $(1.25 \pm 0.76) \times 10^{-4}$ while Jeffrey and Anders (1970) derive $(1.4 \pm 0.4) \times 10^{-4}$ and Black (1972) finds $(1.5 \pm 1.0) \times 10^{-4}$. More recently, Frick and Moniot (1977) claim $({}^3He/{}^4He)_{pre\odot} = (1.588 \pm 0.055) \times 10^{-4}$ and Eberhardt (1978) quotes $({}^3He/{}^4He)_{pre\odot} = (1.46 \pm 0.07) \times 10^{-4}$.

If we average these results (and adopt the "standard" interpretations), the meteoritic and solar wind data suggest a presolar abundance of D *plus* ^{3}He of

$$<(D + {}^3He)/{}^4He>_{pre\odot} = (4.0 \pm 0.2) \times 10^{-4} \qquad (1a)$$

and a presolar ^{3}He abundance of

$$<{}^3He/{}^4He>_{pre\odot} = (1.5 \pm 0.3) \times 10^{-4} \ . \qquad (1b)$$

To obtain an *indirect* estimate of the presolar deuterium abundance, we may subtract (1b) from (1a):

$$<D/^4He>_{pre\odot} = (2.5 \pm 0.5) \times 10^{-4} \ . \tag{1c}$$

5.3.2 Jupiter and Saturn

Observations of molecular lines and bands in the atmospheres of the giant planets provide data on the deuterium abundance. However, it is well known from interstellar studies that chemical and/or physical fractionation effects can distort the "true" isotope ratios. Indeed, it is already known that deuterium is enhanced in water on the Earth and in meteorites (Black 1971, 1972; Geiss and Reeves 1972, 1981). Geiss and Reeves (1981), noting that deuterium is enriched in dark interstellar clouds, caution that the presolar nebula may *not* have provided a fair sample of deuterium in the interstellar medium 4.5 billion years ago. In any case, observations of deuterated molecules in the atmospheres of the giant planets may provide more direct information about the chemistry and physics of the presolar nebula and of planetary atmospheres than about the presolar abundance of deuterium. To minimize the local (i.e., planetary atmosphere) effects of fractionation, it would seem advisable to concentrate on those molecules which contain most of the hydrogen and most of the deuterium. For Jupiter, Trauger et al. (1973) compared lines of HD and H_2. McKellar et al. (1976) used revised laboratory parameters along with the data of Trauger et al. (1973) to derive $(D/H)_{Jup} = (5.6 \pm 1.4) \times 10^{-5}$. Using new data, Trauger et al. (1977) found $(D/H)_{Jup} = (5.1 \pm 0.7) \times 10^{-5}$. In a much more model dependent analysis, Combes and Encrenaz (1979) compared HD with CH_4 and, after correcting for the C/H ratio, derived $(D/H)_{Jup} = (1.3 \pm 1.0) \times 10^{-5}$. In a more recent analysis which changes the C/H ratio, Encrenaz and Combes (1982) find $1.2 \leq 10^5(D/H)_{Jup} \leq 3.1$. In an equally model dependent approach, Kunde et al. (1982) compare CH_3D with CH_4 to derive $2.2 \leq 10^5(D/H)_{Jup} \leq 4.6$.

For Saturn, Encrenaz and Combes (1982) used the HD observations of Macy and Smith (1978), the laboratory data of McKellar et al. (1976), and their data on CH_4 to infer $2 \leq 10^5(D/H)_{Sat} \leq 15$.

In summary, the deuterium abundances derived from observations of Jupiter and Saturn are highly uncertain and very model dependent, and may not even be representative of the presolar nebula or the interstellar medium 4.5 billion years ago. Nonetheless, it is not inconsistent with all the existing data to conclude that the presolar abundance of deuterium (D/H) was in excess of $1\text{–}2 \times 10^{-5}$.

5.3.3 The Sun

It is very difficult to derive accurate abundances from observations of ^{3}He and ^{4}He line intensities in the Sun. From spectroscopic studies of ^{3}He in the wings of ^{4}He in quiescent solar prominences, Hall (1975) found [(D + ^{3}He)/^{4}He]$_{pre\odot}$ = (4 ± 2) × 10^{-4}. This result, although it has a large uncertainty, is in excellent agreement with that inferred from the meteoritic and solar wind data (see § 5.3.1). Heasley and Milkey (1978) measured ^{4}He line intensities in solar prominences and derived (^{4}He/H)$_\odot$ = 0.10 ± 0.025; this corresponds (for $Z_\odot$ = 0.02) to $Y_\odot$ = 0.28 ± 0.05. From a comparison of solar models with observations (solar luminosity, radius, age, etc.), Bahcall et al. (1982) derive an indirect estimate of the solar helium abundance: $Y_\odot$ = 0.25 ± 0.01. Christensen-Dalsgaard and Gough (1980) and Gough (1983) make a detailed comparison of the periods of solar surface oscillations (Grec et al. 1980, 1981) with model predictions to derive $Y_\odot$ = 0.25 ± 0.02.

In the following I will adopt (following YTSSO) for the solar helium abundance (^{4}He/H)$_\odot$ = 0.09 ± 0.01; $Y_\odot$ = 0.26 ± 0.02. Using this ^{4}He abundance and the previously adopted D and ^{3}He ratios (see eq. [1]), we find

$$\left(\frac{D + {}^3He}{H} \right)_{pre\odot} \leq 4.2 \times 10^{-5} \ , \tag{2a}$$

$$\left(\frac{{}^3He}{H} \right)_{pre\odot} \leq 1.8 \times 10^{-5} \ , \tag{2b}$$

$$\left(\frac{D}{H} \right)_{pre\odot} \geq 1.6 \times 10^{-5} \ . \tag{2c}$$

5.4 Stellar and Interstellar Abundance Estimates

5.4.1 Lithium-7

It has already been noted (§ 5.3.1) that the abundance of ^{7}Li derived from meteoritic data (Mason 1979) agrees with that found from interstellar absorption studies (Vanden Bout et al. 1978) and that inferred from spectroscopic studies of Population I stars (Zappala 1972; Duncan 1981). All of these data are consistent with (^{7}Li/H)$_{Pop\ I}$ within a factor of 2 of 1 × 10^{-9}. Since ^{7}Li will be destroyed in some stars (e.g., the Sun) and produced in others (Cameron and Fowler 1971; Scalo 1976; Starrfield et al. 1978), it is very difficult to utilize the Population I abundance of ^{7}Li to infer the primordial abundance. For this reason, the recent observations of Spite and Spite (1982 a, b) of ^{7}Li in the atmospheres of 12 halo and old disk stars (Population II?) are very exciting and extremely important. From their data, the Spites find evidence for an unastrated (despite the presumably extreme ages of the stars in their sample) Population II abundance,

$$\left(\frac{^7\mathrm{Li}}{\mathrm{H}}\right)_{\mathrm{Pop\ II}} = (1.12 \pm 0.38) \times 10^{-10} \ . \tag{3}$$

5.4.2 Deuterium

Interstellar deuterium may be probed by its absorption of the ultraviolet flux from nearby hot stars (York and Rogerson 1976). Such studies have led to estimates of the deuterium abundance in the "local" interstellar medium ($R \lesssim 1$ kpc) (York and Rogerson 1976; Vidal-Madjar et al. 1977; Vidal-Madjar et al. 1983). Although the quoted accuracy in a typical determination of the D/H ratio is good ($\sim\pm25\%$), the current data span the range $(\mathrm{D/H})_{\mathrm{ISM}} = (1/4\text{--}4) \times 10^{-5}$. It is unclear therefore what significance is attached to the "average" value of $<\mathrm{D/H}>_{\mathrm{ISM}} = 2 \times 10^{-5}$ (Bruston et al. 1981). Indeed, only a tiny amount ($\sim$1 part in 10^5) of hydrogen at the "wrong" velocity can mimic deuterium. York (private communication 1983) notes that for the "cleanest" lines of sight the data are consistent with $(\mathrm{D/H})_{\mathrm{ISM}} = 2 \times 10^{-5}$ ($\pm25\%$).

5.4.3 Helium-3

Interstellar helium-3 can be observed through emission of the radio hyperfine line of singly ionized ^{3}He in H II regions. Recent observations by Rood et al. (1983) have detected ^{3}He in the H II regions W43, W51, and W3; the derived abundances cover a wide range: $4 \lesssim 10^5(^3\mathrm{He/H}) \lesssim 20$. For the H II regions W49 and M17, Rood et al. (1983) are only able to set an upper limit: $^3\mathrm{He/H} \lesssim 2 \times 10^{-5}$. Thus, for some H II regions it appears that $(^3\mathrm{He/H})_{\mathrm{ISM}} \gtrsim (^3\mathrm{He/H})_{\mathrm{pre}\odot}$. However, the large uncertainties in both the solar system and H II region abundances preclude our establishing this inequality with much certainty at present.

5.5 Evolution of Deuterium and Helium-3

Virtually all deuterium cycled through stars is destroyed so that it is expected that the primordial abundance exceeds the presolar abundance; the presolar abundance should—in the absence of production of deuterium—be no smaller than that in the present interstellar medium. Deuterium is burned to helium-3, which is more difficult to destroy (D burns at $T \gtrsim 6 \times 10^5$ K, ^{3}He burns at $T \gtrsim 7 \times 10^6$ K). As a result, some of the primordial deuterium will be converted to ^{3}He, which survives to contribute to the present (or presolar) abundance of ^{3}He. To avoid an overabundance of ^{3}He requires that the primordial abundance of D be bounded from above (YTSSO).

Consider the interstellar abundance of deuterium at an arbitrary time in the evolution of the Galaxy and compare it to the primordial abundance. Since deuterium is destroyed in the course of galactic evolution,

$$y_2(t) = f(t)y_{2p} \ . \tag{4}$$

[For convenience we introduce the notation $y_2 \equiv 10^5(\mathrm{D/H})$ and $y_3 \equiv 10^5(^3\mathrm{He/H})$.] In equation (4), $f(t)$ is the "virgin" fraction—the fraction of the interstellar gas, at time t, which has *not* been cycled through stars (i.e., at temperatures in excess of $\sim 6 \times 10^5$ K). Now, low mass stars are net producers of ^{3}He (Iben 1967a, b; Rood 1972), so that the ^{3}He abundance should satisfy

$$y_3(t) \geq f(t)y_{3p} + [1 - f(t)]gy_{23p} \ . \tag{5}$$

In equation (5), $y_{23} \equiv y_2 + y_3$ and g is the fraction of ^{3}He which survives stellar processing; the inequality is to account for the neglect of the "new" ^{3}He produced by low mass stars. Straightforward manipulation of equations (4) and (5) leads to an upper bound on the sum of the primordial abundances of D and ^{3}He

$$y_{23p} \leq y_2(t) + g^{-1}y_3(t) \ . \tag{6}$$

Or

$$y_{23p} \leq y_{23}(t) + (g^{-1} - 1)y_3(t) \ . \tag{6'}$$

To utilize this result to bound y_{23p} from *above* requires that we have reliable *upper* bounds to y_2 and y_3 (or y_{23}). Since the results of Rood et al. (1983) on ^{3}He in H II regions are strongly suggestive of significant stellar contamination of the present interstellar gas, the presolar abundances should—to the extent they are reliable indicators of the abundances in the interstellar medium 4.5 billion years ago—provide the best limits. From equations (2a) and (2b) we will adopt $y_{23\odot} \leq 4.2$ and $y_{3\odot} \leq 1.8$. Finally, we must bound from *below*, g, the surviving fraction of ^{3}He. For stars less massive than 8 $M_\odot$, Iben and

Truran (1978) find $g \geq 0.7$. For higher mass stars ($8 < M/M_\odot < 100$) the models of Dearborn et al. (1978) yield $g \gtrsim 1/4$ (Dearborn 1983); Brunish and Truran (1983) also find $g \gtrsim 1/4$ for $10 < M/M_\odot < 50$. These results for individual stars should be convolved with an IMF to estimate the average g for an entire generation of stars. For a Salpeter (1955) IMF, Brunish and Truran (1983) find $<g> \gtrsim 1/2$. For $g \gtrsim 1/4$–$1/2$ and the solar bounds to y_{23} and y_3, we derive for an upper limit to the sum of the primordial abundances of D and ^{3}He

$$\left(\frac{D + {}^3He}{H} \right)_p \leq (6\text{–}10) \times 10^{-5} \; . \tag{7}$$

In the absence of production of deuterium in the course of galactic evolution, $y_{2p} \geq y_2(t)$. Although the estimates of the interstellar and presolar abundances of deuterium are highly uncertain, the current data suggest $y_{2\,\text{ISM}} \approx y_{2\odot} \gtrsim 1\text{–}2$. In the subsequent discussion it will be assumed that

$$\left(\frac{D}{H} \right)_p \geq (1\text{–}2) \times 10^{-5} \; . \tag{8}$$

5.6 Comparison of Theory and Observation

At this point it is worthwhile comparing the primordial abundances of D, D + ^{3}He, and ^{7}Li—inferred from the observational data—with the predictions of the standard model (see figure 5.1 and table 5.1). We defer for the moment the crucially important comparison of the predicted and observed abundances of ^{4}He.

We have used the solar system data and the interstellar observations of deuterium to derive a lower bound to the primordial abundance. A *lower* bound to $(D/H)_p$ leads to an *upper* bound to the nucleon abundance η. For $(D/H)_p \gtrsim 1(2) \times 10^{-5}$, $\eta_{10} \lesssim 10(7)$. Although $\eta_{10} \approx 10$ is not inconsistent with the deuterium data, such a large nucleon abundance corresponds to a large predicted abundance of lithium-7 ((^{7}Li/H)$_p \approx 9 \times 10^{-10}$). Of course, some ^{7}Li may have been destroyed before the formation of the Population II stars, which the Spites (1982 a, b) have observed. However, in any material in which the ^{7}Li has been destroyed, the deuterium would have been burned away too. Since the present (or presolar) deuterium abundance should be no larger than that at the time the Population II stars were formed, we may require

$$\left(\frac{^{7}\text{Li}}{\text{D}}\right)_{p} \lesssim \left(\frac{^{7}\text{Li}}{\text{H}}\right)_{\text{Pop II}} \left[\left(\frac{\text{D}}{\text{H}}\right)_{0}\right]^{-1} \lesssim 2 \times 10^{-5} \ . \tag{9}$$

From table 5.1 it follows that this constraint requires $\eta_{10} \lesssim 7$. It must be reemphasized, however, that the important observations of the Spites (1982a, b) should receive further confirmation; also, recall that the ^{7}Li abundance predicted by the standard model is uncertain by a factor of 2. We therefore conclude that, whereas the current data suggest $\eta_{10} \lesssim 7$, a firm upper bound is $\eta_{10} \lesssim 10$.

If the Spites' (1982a, b) observations of ^{7}Li in Population II stars provide a good estimate of the primordial abundance, then $(^{7}\text{Li/H})_{p} \approx (1 \pm 1/2) \times 10^{-10}$. The standard model will account for this abundance provided that the nucleon abundance lies in the range $2 \lesssim \eta_{10} \lesssim 5$.

Finally, if we employ the upper bound to the sum of the primordial abundances of D and ^{3}He derived in the previous section, we obtain a lower bound to the nucleon abundance. For $[(\text{D} + {}^{3}\text{He})/\text{H}]_{p} \lesssim (6\text{--}10) \times 10^{-5}$, we find $\eta_{10} \gtrsim 3\text{--}4$.

To summarize, the primordial abundances of D, ^{3}He, and ^{7}Li inferred from current observational data are in accord with the predictions of the standard model provided that the nucleon abundance is in the range

$$3\text{--}4 \lesssim \eta_{10} \lesssim 7\text{--}10 \ . \tag{10}$$

To establish the overall consistency of the standard model (or to discover its inconsistency) we must now compare the predicted abundance of primordial helium-4 with that inferred from the observational data. Note that for $\tau_{1/2} \geq 10.4$ min., $N_{\nu} = 3$, and $\eta_{10} \geq 3$: $Y_{p} \geq 0.240$.

5.7 Primordial Abundance of Helium-4

The most accurate ^{4}He abundances are derived from studies of the emission lines in galactic and extragalactic H II regions, and I will limit my discussion to such data. The entire subject of primordial helium has been considered in a recent ESO workshop, to which the interested reader is referred for the most comprehensive and up-to-date review of the observational situation. The issue of primordial helium has also been reviewed by Pagel (1982) and by YTSSO. Here, I will limit my discussion to the most recent high quality observations.

All the data at present are consistent with a primordial ^{4}He abundance (by mass) $Y_p = 0.24 \pm 0.02$. I am aware of no observer who would disagree with this statement. For ^{4}He in this range, the standard model (with 3–4 $\lesssim \eta_{10} \lesssim$ 7–10) is completely consistent. To test the standard model requires that the uncertainty in Y_p be reduced, a goal which is very difficult to achieve. Higher accuracy requires care with observations and data reduction which is unusual in nebular spectroscopy; for excellent discussions of the problems, perils, and pitfalls, the reader is urged to consult Kunth and Sargent (1983) and Kinman and Davidson (1983).

Even with the highest quality observational data and with the most careful data reduction, one is still a long way from the primordial abundance of helium. Recombination lines of ^{4}He$^+$ (and, occasionally, ^{4}He^{++}) are observed; a "correction" for neutral helium must be made. Rayo et al. (1982) use the models of Stasinska (1980) to correct for neutral helium but find that some of their observed H II regions are fit by none of her models. Kunth and Sargent (1983) use an ionization correction ($i_{CF}y^+ = y^0 + y^+$; $y^+ = $ He$^+$/H$^+$, $y^0 = $ He^0/H$^+$) deduced from the models of Peimbert et al. (1974):

$$i_{CF}^{-1} = 1 - 0.25(O^+/O) \ . \tag{11}$$

Lequeux et al. (1979) employ a different ionization correction factor:

$$i_{CF}^{-1} = 1 - 0.15(O^+/O) - 0.85(S^+/S) \ . \tag{12}$$

For comparison, the fact that the ionization potential of S II is close to that of He I (23.4 eV versus 24.6 eV) suggests that

$$i_{CF} \approx 1 + S^+/S^{++} \ . \tag{13}$$

Even if, in individual cases, the ionization correction is small, the *uncertainty* in that correction must not be ignored in estimating the total uncertainty in the derived abundance.

Finally, since stars produce ^{4}He along with the heavier elements, the truly primordial abundance will be contaminated by stellar production. A correction must be made therefore for the amount of ^{4}He (Y_{ev}) produced in the course of galactic evolution. Once again, the *uncertainty* in this correction adds to the uncertainty in the derived primordial abundance. For example, Lequeux et al. (1979) find $\Delta Y/\Delta Z = 1.7 \pm 0.9$ (Z is the heavy element abundance by mass) whereas Kunth and Sargent's (1983) data imply $\Delta Y/\Delta Z \leq 1.3 \pm 3.6$. Furthermore, since other heavy elements may not increase like oxygen, it may be more meaningful to compare ΔY with $\Delta Z(O)$ or $\Delta(O/H)$ (Rayo et al. 1982). In terms of the helium and heavy element mass fractions Y and Z (and $y = He/H$) we have

$$Y = \frac{4y(1 - Z)}{(1 + 4y)} \, , \tag{14a}$$

$$Z(O) = 16 \left(\frac{O}{H} \right) \frac{(1 - Z)}{(1 + 4y)} \, , \tag{14b}$$

so that

$$\Delta y = 4 \left(\frac{\Delta Y}{\Delta Z} \right) \left(\frac{Z}{Z(O)} \right) \Delta \left(\frac{O}{H} \right) \, . \tag{14c}$$

For $\Delta Y/\Delta Z \approx 2 \pm 1$ and $Z \approx 2.2 \, Z(O)$, $\Delta y \approx (18 \pm 9)\Delta(O/H)$.

Let us combine these two effects—ionization correction and evolution—to obtain a general result for the primordial abundance of ^{4}He. Analysis of the spectroscopic data will lead to the observed ratio of He^+ to H^+:

$$y^+_{obs} \equiv y_1 \pm \Delta y_1 \, . \tag{15}$$

For excellent data and careful analysis (Kunth and Sargent 1983), $\Delta y_1 \approx 0.003$–0.006. The neutral helium fraction (He^0/H^+) must be inferred from models or from the ionization correction factor:

$$y^0 = (i_{CF} - 1)y^+_{obs} \equiv y_2 \pm \Delta y_2 \, . \tag{16}$$

For low metal abundance H II regions (e.g., Kunth and Sargent 1983), $i_{CF} \approx 1.05 \pm 0.05$ so that $y_2 \approx \Delta y_2 \approx 0.004$. Finally, the stellar-produced ^{4}He should be subtracted:

$$\Delta y_{ev} \approx (18 \pm 9)\Delta(O/H) \equiv y_3 \pm \Delta y_3 \, . \tag{17}$$

For $\Delta(O/H) \approx 1 \times 10^{-4}$, $y_3 \approx 0.002$ and $\Delta y_3 \approx 0.001$. Combining equations (15)–(17) we derive the primordial value

$$y_p = y^+_{obs} + y^0 - \Delta y_{ev} = (y_1 + y_2 - y_3) \pm (\Delta y_1 + \Delta y_2 + \Delta y_3) \, , \tag{18a}$$

$$y_p \approx (y_1 + 0.002) \pm (0.008\text{–}0.011) \, . \tag{18b}$$

For $y_1 \approx 0.08$, typical of H II regions with $O/H \approx (1/2\text{–}2) \times 10^{-4}$, $Y_p \approx 3y_p$ so that

$$Y_p \approx 3y_p \approx (Y_1 + 0.006) \pm (0.02\text{–}0.03) \, . \tag{19}$$

It is the last term in equation (19) which is important; even the best *individual* H II regions studied should not yield a primordial mass fraction of ^{4}He more accurate than ± 0.02.

Any attempt therefore to derive the primordial abundance of helium from the observation of one H II region should be scrutinized with extreme care. The recent work of Rayo et al. (1982) on the H II region NGC 5471 in the galaxy M101 is an example of just such an attempt. For this object, Rayo et al. (1982) find $y^+ = 0.0721 \pm 0.0009$. As Kunth and Sargent (1983) have noted, this quoted uncertainty—a factor of 4 smaller than that in other very high quality work—seems artificially small. Furthermore, using the models of Stasinska (1980), Rayo et al. (1982) claim that no correction for neutral helium is required. However, using their data and the ionization correction factors in equations (11)–(13) we would find $1.072 \leq i_{CF} \leq 1.126$, so that $0.005 \lesssim y^0 \lesssim 0.009$. It should be noted that in his study of NGC 5471 Smith (1975) found $1.10 \lesssim i_{CF} \lesssim 1.16$. Finally, Rayo et al. (1982) correct for stellar-produced helium by using the result of Lequeux et al. (1979): $\Delta Y / \Delta Z(\text{O}) = 3.8$ (no uncertainty is attached to this evolutionary correction). If we assume, following Lequeux et al. (1979), that $\Delta Y / \Delta Z(\text{O}) = 3.8 \pm 2.0$ and adopt from Rayo et al. (1982) $\Delta(\text{O}/\text{H}) = (1.55 \pm 0.15) \times 10^{-4}$, we obtain $\Delta y_{\text{ev}} = 0.0024 \pm 0.0015$. If we now combine all these terms (y^+, y^0, Δy_{ev}), we derive

$$y_p = 0.0767 \pm 0.0044; \quad Y_p = 0.235 \pm 0.010 \ . \tag{20}$$

It is not the central values themselves in equation (20) which are significant but the uncertainties, which are larger than those claimed by Rayo et al. (1982) by a factor of 5. In obtaining the primordial abundance from one H II region, buyer beware!

In their recent work, Kunth and Sargent (1983) have obtained data on 13 low metal abundance $(0.3 \lesssim 10^4(\text{O}/\text{H}) \lesssim 2.0)$ H II regions. They correct their observations of y^+ (and, where appropriate, y^{++}) for neutral helium via equation (11). They find no evidence for a y versus O/H trend and conclude $Y_p = 0.245 \pm 0.003$ ($y_p = 0.0811 \pm 0.0013$). Some comments are in order here too. The bulk of the Kunth-Sargent data spans a range in oxygen abundance of only $\Delta(\text{O}/\text{H}) \approx 1 \times 10^{-4}$. In that case, even if $\Delta Y / \Delta Z \approx 2 \pm 1$, we would expect $\Delta y_{\text{ev}} \approx 0.002 \pm 0.001$. It is therefore not surprising that Kunth and Sargent (1983) find no strong evidence for a Y versus Z trend. Indeed, they do find that their data are not inconsistent with $\Delta Y / \Delta Z = 1.3 \pm 3.6$. Regarding the ionization correction, Kunth and Sargent (1983) do comment that its uncertainty is a source of systematic error, but they do not include an estimate of that error in their overall uncertainties. If we assume an uncertainty $\Delta i_{CF} \approx \pm 0.05$ for each H II region and make an evolutionary correction according to Lequeux et al. (1979) ($\Delta y_{ev} \approx 0.0018 \pm 0.0009$), we may reinterpret the Kunth-Sargent results as

$$y_p = 0.0793 \pm 0.0033; \; Y_p = 0.241 \pm 0.008 \; . \tag{21}$$

With due account of the uncertainties, the results of Kunth and Sargent (eq. [21]) and Lequeux et al. ($Y_p = 0.233 \pm 0.005$) are entirely consistent.

At the beginning of this section, I commented that all the data at present are consistent with $Y_p = 0.24 \pm 0.02$. The recent very high quality data have not changed this conclusion except, perhaps, to reduce the uncertainty by a factor of 2.

At present, I conclude from the existing data that $Y_p = 0.24 \pm 0.01$; a third significant figure in either $<Y_p>$ or ΔY_p would be illusory. The predicted abundance (for $N_\nu = 3$ and $\tau_{1/2} = 10.6$ min.) is shown in table 5.1. Recall that the uncertainty in the neutron half-life ($\Delta \tau_{1/2} \approx \pm 0.2$ min.) corresponds to an uncertainty in the predicted abundance of $\Delta Y_p \approx \pm 0.003$. It is clear from table 5.1 that the predictions of the standard model are consistent with the observational data for $\eta_{10} \lesssim 8$–10 (remember, we are not taking the third significant figure seriously so that $0.252 \approx 0.25$). For the nucleon abundance in the range 3–$4 \lesssim \eta_{10} \lesssim 7$–$10$, there is complete consistency between the predictions of the standard model and the observed abundances of D, ^{3}He, ^{4}He, and ^{7}Li.

5.8 Conclusions

A detailed comparison of theory and observation leaves the standard model smelling like a rose! There is at present complete consistency between the abundances of D, ^{3}He, ^{4}He, and ^{7}Li predicted by primordial nucleosynthesis in the standard, hot big bang model and the primordial abundances of those elements inferred from current data. Lest we be accused of complacency (which, inevitably, we will be), we should remember that the uncertainties in the present abundances are still quite large and the evolutionary models used to infer the primordial abundances are crude. Lest the current concordance be a delusion, we must continue to refine our data and strive to build better models. It would be churlish, however, to ignore the current good agreement and to fail to draw the consequences of interest to cosmology and particle physics. I conclude this paper therefore by listing those results.

5.8.1 Nucleons Fail to Close the Universe

If the nucleon density were equal to the critical density, then the nucleon to photon ratio would exceed $\eta_{10} \gtrsim 52$ (YTSSO). For such a large value of η, the predicted abundance of deuterium would be two orders of magnitude lower than that observed, the predicted abundance of lithium would exceed that observed by a factor of 10–100, and the helium abundance would exceed $Y_p \gtrsim 0.27$.

5.8.2 A Nucleon Dominated Universe Is a Low Density Universe

For $\eta_{10} \lesssim 7$–10, nucleons account for less than 14–19% of the critical density $(\Omega_N \lesssim 0.14$–0.19). Nucleons are, however, capable of accounting for the mass in galaxies $(\Omega_N \gtrsim 0.01)$.

5.8.3 At Most, Only One "Extra" Neutrino Is Permitted

The presence of an additional, light neutrino $(N_\nu = 4)$ would increase the predicted helium abundance by $\Delta Y_p \approx 0.01$ so that consistency with $Y_p \lesssim 0.25$ would require $\eta_{10} \lesssim 3$ (see table 5.1). The current data therefore permit no more than one extra neutrino (or equivalent "ino").

Acknowledgments

All of the material in this paper is taken directly from work done in collaboration with K. A. Olive, D. N. Schramm, M. S. Turner, and J. Yang (YTSSO). For further details and a more extensive list of references the interested reader is urged to consult that paper. Since YTSSO was written, I have learned a great deal from conversations with and communications from K. Davidson, L. Hobbs, H. Reeves, R. T. Rood, and P. A. Shaver. This work is supported at Bartol by DOE DE-ACO2-78ER-05007.

References

Anders, E., Heymann, D., and Mazor, E. 1970, *Geochim. Cosmochim. Acta,* **34,** 127.

Bahcall, J. N.; Huebner, W. F.; Lubow, S. H.; Parker, P. D.; and Ulrich, R. K. 1982, *Rev. Mod. Phys.,* **54,** 767.

Beaudet, G., and Reeves, H. 1983, in *Proc. of the ESO Workshop on Primordial Helium,* ed. P. A. Shaver, D. Kunth, and K. Kjar, p. 53.

Black, D. C. 1971, *Nature Phys. Sci.,* **234,** 148.

——. 1972, *Geochim. Cosmochim. Acta,* **36,** 347.

Brunish, W., and Truran, J. W. 1983, in preparation.

Bruston, P., Audouze, J., Vidal-Madjar, A., and Laurent, C. 1981, *Ap. J.,* **243,** 161.

Cameron, A. G. W., and Fowler, W. A. 1971, *Ap. J.,* **164,** 111.

Christensen-Dalsgaard, J., and Gough, D. O. 1980, *Nature,* **288,** 544.

Combes, M., and Encrenaz, T. 1979, *Icarus,* **39,** 1.

Dearborn, D. S. P. 1983, private communication.

Dearborn, D. S. P., Blake, J. B., Hainebach, K. L., and Schramm, D. N. 1978, *Ap. J.,* **223,** 552.

Dicus, D. A., Kolb, E. W., Gleeson, A. M., Sudarshan, E. C. G., Teplitz, V. L., and Turner, M. S. 1982, *Phys. Rev.,* **D26,** 2694.

Duncan, D. K. 1981, *Ap. J.,* **248,** 651.

Eberhardt, P. 1978, *Proc. 9th Lunar Sci. Conf.,* p. 1027.

Encrenaz, T., and Combes, M. 1982, *Icarus,* **52,** 54.

Frick, U., and Moniot, R. K. 1977, *Proc. 8th Lunar Sci. Conf.,* p. 229.

Geiss, J., and Reeves, H. 1972, *Astr. Ap.,* **18,** 126.

——. 1981, *Astr. Ap.,* **93,** 189.

Gough, D. O. 1983, in *Proc. of the ESO Workshop on Primordial Helium,* ed. P. A. Shaver, D. Kunth, and K. Kjar, p. 117.

Grec, G., Fossat, E., and Pomerantz, M. A. 1980, *Nature,* **288,** 541.

——. 1983, *Solar Phys.,* **82,** 55.

Hall, D. N. B. 1975, *Ap. J.,* **197,** 509.

Heasley, J. N., and Milkey, R. W. 1978, *Ap. J.,* **243,** 127.

Iben, I. 1967a, *Ap. J.,* **147,** 624.

——. 1967b, *Ap. J.,* **147,** 650.

Iben, I., and Truran, J. W. 1978, *Ap. J.,* **220,** 980.

Jeffrey, P. M., and Anders, E. 1970, *Geochim. Cosmochim. Acta,* **34,** 1175.

Kinman, T. D., and Davidson, K. 1983, submitted to *Ap. J.*

Kunde, V., Hanel, R., Maguire, W., Gautier, D., Baluteau, J. P., Marten, A., Chedin, A., Husson, N., and Scott, N. 1982, *Ap. J.,* **263,** 443.

Kunth, D., and Sargent, W. L. W. 1983, submitted to *Ap. J.*

Lequeux, J., Peimbert, M., Rayo, J. F., Serrano, A., and Torres-Peimbert, S. 1979, *Astr. Ap.*, **80**, 155.

Macy, W., Jr., and Smith, W. H. 1978, *Ap. J. (Letters)*, **222**, L73.

Mason, B. 1979, "Cosmochemistry, Part I: Meteorites," in chapter B in *Data of Geochemistry*, 6th ed., ed. M. Fleischer (Washington, D. C.: Government Printing Office).

McKellar, A. R. W., Goetz, W., and Ramsey, D. A. 1976, *Ap. J.*, **207**, 663.

Pagel, B. E. J. 1982, *Phil. Trans. Roy. Soc.*, **A307**, 19.

Peimbert, M., Rodrigues, L. F., and Torres-Peimbert, S. 1974, *Rev. Mex. Astr. Ap.*, **1**, 129.

Rayo, J., Peimbert, M., and Torres-Peimbert, S. 1982, *Ap. J.*, **255**, 1.

Rood, R. T. 1972, *Ap. J.*, **177**, 681.

Rood, R. T., Wilson, T. L., and Bania, T. M. 1983, submitted to *Ap. J.*

Salpeter, E. E. 1955, *Ap. J.*, **121**, 161.

Scalo, J. M. 1976, *Ap. J.*, **206**, 795.

Smith, H. E. 1975, *Ap. J.*, **199**, 591.

Spite, F., and Spite, M. 1982a, *Astr. Ap.*, **115**, 357.

——. 1982b, *Nature*, **297**, 483.

Starrfield, S., Truran, J. W., Sparks, W. M., and Arnould, M. 1978, *Ap. J.*, **222**, 600.

Stasinska, G. 1980, *Astr. Ap.*, **84**, 320.

Trauger, J. T., Roesler, F. L., Carleton, N. P., and Traub, W. A. 1973, *Ap. J. (Letters)*, **184**, L137.

Trauger, J. T., Roesler, F. L., and Mickelson, M. E. 1977, *Bull. AAS*, **9**, 516.

Vanden Bout, P. A., Snell, R. L., Vogt, S. S., and Tull, R. G. 1978, *Ap. J.*, **221**, 598.

Vidal-Madjar, A.; Laurent, C.; Bonnet, R. M.; and York, D. G. 1977, *Ap. J.*, **211**, 91.

Vidal-Madjar, A., Laurent, C., Gry, C., Bruston, P., Ferlet, R., and York, D. G. 1983, *Astr. Ap.*, **120**, 58.

Yang, J., Turner, M. S., Steigman, G., Schramm, D. N., and Olive, K. A. 1984, *Ap. J.*, in press (YTSSO).

York, D. G., and Rogerson, J. B., Jr. 1976, *Ap. J.*, **203**, 378.

Zappala, R. A. 1972, *Ap. J.*, **172**, 57.

6

Galactic Chemical Evolution and Nucleocosmochronology: A Standard Model

Donald D. Clayton

Abstract

I present a new analytic model of the chemical evolution of the galaxy. Explicit solutions for a gas mass, star mass, metallicity, and radiochronometers are obtained for models in which the galactic infall can be parameterized as $f(t) = kM_{G0}/\Delta[(t + \Delta)/\Delta]^{k-1}e^{-\omega t}$, where $k =$ integer, $\Delta =$ parameter, and M_{G0} is initial disk mass, and for which the star formation rate is proportional to the gas mass. Because all physically interesting observables have a simple explicit dependence on the parameters which themselves map a wide space of physical possibilities, I suggest adopting this model as a reference standard for studies of chemical evolution.

6.1 Introduction

In this work I present a new analytic model of the chemical evolution of the galaxy. I even suggest that it makes a good "standard model" for galactic evolution, because known properties of chemical evolution are well modeled by it. In setting out upon this task I held it as desirable that the abundances of radioactive chronometers be expressible analytically, so that the cosmochronometers could display explicit dependence upon the properties of the total model. The gas mass $M_G(t)$, the star mass $S(t)$, and the metallicity $Z(t)$ are also expressed analytically. With such a class of models, the major tests of chemical evolution are quickly evaluated, so that their dependence upon the model histories can be more instructive than are computer printouts. In this sense a good "standard model" of chemical evolution can play the same useful role as does a standard analytic model of a star, or of other astronomical objects; namely in providing an easily understandable frame of reference within which more specific and special results can be comprehended.

In constructing this model I have been guided by a few astrophysical biases in addition to the need to see the nuclear cosmochronology explicitly. Despite much discussion of more complicated models, I find it most natural to assume that the initial mass function is time independent, and that the star formation rate $\psi(t)$ depends linearly upon the gas mass $M_G(t)$. It is also natural to assume that the disk component of the galaxy grew as matter fell upon it during the galaxy's maturation, with the total mass of the disk growing to many times its initial value (defined as the value where star formation began in the disk). Because the infall, which I call $f(t)$ following Tinsley's (1980) excellent description of galactic evolution, has probably decreased in time, and may practically have ceased today, I choose an infall rate having that property. From this it follows that the gas mass $M_G(t)$, and hence the star formation rate $\psi(t)$, first grows as matter falls onto the disk, and later decreases as the infall abates. I will also employ the instantaneous recycling approximation; namely that short-lived massive stars so dominate the mass recycling through stars that the time delay between star birth and ejection can be neglected. Many studies have confirmed the general validity of this approximation (e.g., Tinsley 1980 and many references therein). This simplification allows the ejecta to be written $E(t) = R\psi(t)$, where R is the return fraction.

Before turning to a class of solvable models incorporating these astrophysical biases, I must add that my reconsideration of this old and important problem, on which so much has been done by so many (e.g., Tinsley 1980 and references therein), was stimulated directly by the discovery (Mahoney et al. 1982) of live ^{26}Al in the interstellar medium and by uncertainty about its implications for models of the chemical evolution of the galaxy. My analysis (Clayton 1984) of that situation has led me to reconsider the allowable expectations for the abundances of all radioactive nuclei within the interstellar medium; and I suspect that the family of models which I have discovered probably maps the range of physically realistic models.

6.2 Mathematical Linear Model

By the linear model I mean that the star formation rate $\psi(t)$ is taken to be proportional to the gas mass $M_G(t)$, which may itself be thought of as either in the total disk or in a solar annulus. So with a mass fraction R

returned instantaneously, the equation (following Tinsley's notation)

$$\frac{dM_G}{dt} = -\psi(1 - R) + f \tag{1}$$

can be simplified greatly by setting

$$\psi(1 - R) = \omega M_G , \tag{2}$$

where ω is a constant by the assumption of linearity. Where there is no infall ($f = 0$), ω would equal the rate of exponential decrease in the gas mass as it turns into stars. That simple closed exponential model is unsatisfactory, so with $f(t) \neq 0$ the function M_G will not be exponential; nonetheless, the constancy of ω remains reasonable as long as one expects the rate of star formation to be proportional to the mass of gas available for forming stars. I find this to be a very natural assumption even though many have speculated, following Schmidt (1959, 1963), that a steeper nonlinear dependence upon M_G is called for. Keeping the linear assumption, the solution of (1) is

$$M_G(t) = M_{G0} \exp\left[-\omega t + \int_0^t \frac{f(t')}{M_G(t')} \, dt' \right] \tag{3}$$

which shows explicitly how the gas mass is rendered nonexponential by the infall. The ratio $f(t)/M_G(t) \equiv \omega_f(t)$ is the instantaneous rate at which the current $M_G(t)$ is being replenished by the current infall $f(t)$. Because it appears so often in the mathematical analysis, it is convenient to define a dimensionless function of time

$$\theta(t) = \int_0^t \omega_f(t') \, dt' = \int_0^t f(t')/M_G(t') \, dt' \tag{4}$$

where the angular mnemonic $\theta = \bar{\omega}_f t$ allows us to understand θ as being the number of times that the interstellar medium has been cycled by infall. Then $M_G(t) = M_{G0} \exp(-\omega t + \theta(t))$ gives the time dependence of the gas.

Lynden-Bell (1975) has emphasized the importance of having a nonmonotonic function $M_G(t)$. Under heavy infall $M_G(t)$ grows at first, but because it is initially small, the number of low-Z stars is not severe since $Z(t)$ grows quickly, so that the G-dwarf problem is circumvented. Larson (1972) had first made the same physical point by advancing a simple model in which both $M_G(t)$ and $f(t)$ were constants in time. Equation (3) shows clearly that $M_G(t)$ first increases if $\theta(t)$ initially increases faster than ωt. The model that I will construct has that property.

The mass of stars $S(t)$ is given by the integral over time of $\psi(1 - R)$, which by (2) becomes

$$S(t) = \omega M_{G0} \int_0^t \exp(-\omega t' + \theta(t')) \, dt' \quad . \tag{5}$$

Observationally, this function $S(t)$ assumes importance in studying the distribution of stars with age or, more easily, with metallicity.

The equation for the mass fraction $Z(t)$ of a primary nucleosynthesis product is approached through the total mass of element Z in the gas (e.g., Tinsley 1980, eq. [3.18]):

$$\frac{d}{dt}(ZM_G) = -Z(1 - R)\psi + y_z(1 - R)\psi + Z_f f - \lambda ZM_G \quad , \tag{6}$$

where λ is the radioactive decay rate of species Z, where y_z is the "yield" of element Z per unit increase of the mass of stars [remembering that $\psi(1 - R) \, dt$ is the increase in star mass during dt], and where Z_f is the mass fraction of element Z in the infalling matter. For later evaluation, I will recommend taking $Z_f = 0$, at least until that time that astrophysical arguments demand a significant Z_f in the infalling matter. Using (1) in the identity $M_G dZ/dt = d(ZM_G)/dt - Z \, dM_G/dt$ yields the equation for the gas abundance

$$M_G dZ/dt = y_z(1 - R)\psi - (Z - Z_f)f - \lambda M_G Z \tag{7}$$

$$= y_z \omega M_G - (Z - Z_f)f - \lambda M_G Z \quad , \tag{8}$$

where (7) is general and (8) applies to the linear model to be developed herein. Clearly (8) can also be written

$$dZ/dt = (y_z \omega - \lambda Z) - (Z - Z_f) \, f/M_G \tag{9}$$

$$= (y_z \omega - \lambda Z) - (Z - Z_f) \, d\theta/dt \quad . \tag{10}$$

This equation has a very simple solution that appears to have not been previously described:

$$(Z - Z_f) = e^{-\lambda t - \theta}\left[(y_z \omega - \lambda Z_f) \int_0^t e^{\lambda t' + \theta(t')} dt' + Z_0 - Z_f\right] \quad . \tag{11}$$

The infall Z_f is restricted to a constant in this solution, which is not very reasonable except in the special case when it is so small that it essentially vanishes. Thus I will take metal-poor infall for an attempt at a simple standard model. Setting $Z_f = 0$ is physically reasonable if halo stars were not able to appreciably enrich the infalling matter. The initial abundance in the disk Z_0 will also be set to zero in the first attempt. And to determine the growth of stable metallicity one simply sets the radioactive decay rate $\lambda = 0$.

Armed with these results of the linear theory, namely equations (3), (5), and (11), we can search for physically motivated analytic solutions. They require that $\theta(t)$ have a functional form that is both physically meaningful and manageable in equations (5) and (11).

6.3 The Standard Model

The infall function $f(t)$ is free, unrestricted by the equations of chemical evolution, although, of course, it will be determined by the correct dynamic theory of galaxy formation. I here suggest, as a parameterized model, the form

$$\frac{d\theta}{dt} \equiv \omega_f \equiv f(t)/M_G(t) = \frac{k}{t + \Delta} \ , \tag{12}$$

where k is an arbitrary positive integer (for mathematical simplicity) and Δ is an arbitrary parameter. My reasons for choosing this form to define a model are really several: (1) this rate ω_f to replace M_G by infall decreases with time; (2) the absolute infall rate $f(t)$ peaks at an earlier time than does $M_G(t)$; (3) the free choice of two parameters, k and Δ, allows adequate form-fitting to ω_f; and (4) the solutions can be evaluated analytically.

It is really reason (4) that motivates me, because a "standard model" must, if the name is to have useful meaning, be subject to quick interpretation. With that in mind, notice the simplicity of the results. From (4),

$$\theta(t) = \int_0^t f(t')/M_G(t') \ dt' = \ln\left(\frac{t + \Delta}{\Delta}\right)^k \ , \tag{13}$$

whereupon the gas mass

$$M_G(t) = M_{G0}\left(\frac{t + \Delta}{\Delta}\right)^k e^{-\omega t} \tag{14}$$

and the infall rate

$$f(t) = \frac{k}{t + \Delta} \ M_G(t) = \frac{kM_{G0}}{\Delta}\left(\frac{t + \Delta}{\Delta}\right)^{k-1} e^{-\omega t} \tag{15}$$

and the total mass of stars from (5)

$$S(t) = \omega M_{G0} I_k(t,-\omega) \ , \tag{16}$$

where we define a simple integral that is of repeated use in this model and is very easy to evaluate:

$$I_k(t,-\omega) \equiv \int_0^t \left(\frac{t' + \Delta}{\Delta}\right)^k e^{-\omega t'} \, dt' \tag{17}$$

is a simple function of t that depends upon the parameters, k, Δ, and ω. Furthermore, the mass fraction of a primary nucleosynthesis product can also be evaluated as

$$(Z - Z_f) = (y_z\omega - \lambda Z_f)e^{-\lambda t}\left(\frac{t + \Delta}{\Delta}\right)^{-k} \int_0^t \left(\frac{t' + \Delta}{\Delta}\right)^k e^{\lambda t'} \, dt'$$

$$= (y_z\omega - \lambda Z_f)e^{-\lambda t}\left(\frac{\Delta}{t + \Delta}\right)^k I_k(t,\lambda) \ , \tag{18}$$

where the integral $I_k(t,\lambda)$ is the same function as the integral $I_k(t,-\omega)$ that occurs in $S(t)$ except that the parameter $(-\omega)$ has been replaced by the parameter $(+\lambda)$—hence the notation showing explicitly the parameter.

For the case of a stable nucleus, the metallicity reduces to

$$(Z - Z_f) = [y_z\omega\Delta/(k + 1)]\left[\frac{t + \Delta}{\Delta} - \left(\frac{t + \Delta}{\Delta}\right)^{-k}\right] \ . \tag{19}$$

The total mass of the galactic disk at any time is

$$M_{\text{tot}} = M_G + S = M_{G0}\left[\left(\frac{t + \Delta}{\Delta}\right)^k e^{-\omega t} + \omega I_k(t,-\omega)\right] \ . \tag{20}$$

These results specify virtually everything of interest to such models as simple functions of time that can even be evaluated on a hand calculator. The integrals $I_k(t,\lambda)$ are easily evaluated as

$$k = 0: I_0 = (e^{\lambda t} - 1)/\lambda \ ,$$

$$k = 1: I_1 = \frac{1}{\Delta}\left\{e^{\lambda t}\left[\frac{t + \Delta}{\lambda} - \frac{1}{\lambda^2}\right] - \left[\frac{\Delta}{\lambda} - \frac{1}{\lambda^2}\right]\right\} \ ,$$

$$k = 2: I_2 = \frac{1}{\Delta^2}\left\{e^{\lambda t}\left[\frac{(t + \Delta)^2}{\lambda} - \frac{2(t + \Delta)}{\lambda^2} + \frac{2}{\lambda^3}\right] - \left[\frac{\Delta^2}{\lambda} - \frac{2\Delta}{\lambda^2} + \frac{2}{\lambda^3}\right]\right\} \ ,$$

$$k = 3: I_3 = \frac{1}{\Delta^3}\left\{e^{\lambda t}\left[\frac{(t + \Delta)^3}{\lambda} - \frac{3(t + \Delta)^2}{\lambda^2} + \frac{6(t + \Delta)}{\lambda^3} - \frac{6}{\lambda^4}\right]\right.$$

$$-\left[\frac{\Delta^3}{\lambda} - \frac{3\Delta^2}{\lambda^2} + \frac{6\Delta}{\lambda^3} - \frac{6}{\lambda^4}\right]\right\} \ ,$$

and in general

$$I_k = \Delta^{-k}\left\{e^{\lambda t}\left[\frac{(t+\Delta)^k}{\lambda} - \frac{k(t+\Delta)^{k-1}}{\lambda^2} + \frac{k(k-1)(t+\Delta)^{k-2}}{\lambda^3} - \cdots\right.\right.$$

$$\left. + (-1)^{k-1}\frac{k!(t+\Delta)}{\lambda^k} + (-1)^k\frac{k!}{\lambda^{k+1}}\right] - \left[\frac{\Delta^k}{\lambda} - \frac{k\Delta^{k-1}}{\lambda^2} + \frac{k(k-1)\Delta^{k-2}}{\lambda^3} - \cdots\right.$$

$$\left.\left. + (-1)^{k-1}\frac{k!\Delta}{\lambda^k} + (-1)^k\frac{k!}{\lambda^{k+1}}\right]\right\}$$

For arbitrary values of k one can also generate I_k from I_{k-1} by the recursion

$$I_k = \lambda^{-1}\left\{\left[\left(\frac{t+\Delta}{\Delta}\right)^k e^{\lambda t} - 1 - \frac{k}{\Delta}I_{k-1}\right]\right\} \ .$$

For the similar integral $I_k(t,-\omega)$ occurring in (5) and elsewhere one need only replace λ by $-\omega$. Because these functions are products of simple polynomials and exponentials, their evaluation is numerically easy. In presenting the first results of this model, I have been content to use $k = 1$, 2, or 3. For the display and discussion of numerical results I will express time in units of 10^9 years.

6.3.1 Gas Mass $M_G(t)$

As a function of time the gas mass $M_G(t)$ reaches a maximum at $t(\max M_G) = (k/\omega) - \Delta$, as is easily seen by differentiating (14). At earlier times the rate of star formation is also increasing in proportion to the disk gas mass. This feature has the useful result that the number of stars born very early is small. In figure 6.1, I display two examples of the function $M_G(t)$ for the particular choice $k = 1$. For $\Delta = 0.1$ the gas $M_G(t)$ rises to a pronounced maximum (at $t = 2.4$ for the case shown $\omega = 0.4$) whereas for $\Delta = 4$ the maximum is at $t = 0$ (because $k/\omega - \Delta$ is in that case negative). Even these two examples make it evident that a very wide range of shapes for $M_G(t)$ is possible, although they all have a single maximum. The following rules of thumb for $M_G(t)$ are almost self-evident:

(1) increasing k moves $t(\max M_G)$ to larger t;

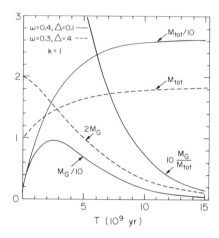

Fig. 6.1. Two contrasting examples of the behavior of the gas mass $M_G(t)$ for models having $k = 1$ are displayed (with scale changes as noted) as multiples of the initial mass M_{G0}. Models with small Δ (*solid lines*) have a maximum before declining because the early infall is large, whereas those having large Δ (*dashed lines*) decrease monotonically owing to the small early infall. The infall itself (not shown) is exponential for $k = 1$ models. Also shown are the total disk masses in units of M_{G0}, showing that the small Δ cases have a greater growth factor. Models having $k > 1$ are qualitatively similar.

(2) increasing ω moves $t(\max M_G)$ to smaller t;

(3) increasing Δ moves $t(\max M_G)$ to smaller t, being at $t = 0$ if $\Delta > k/\omega$;

(4) for a chosen $t(\max M_G)$, the relative height $M_G(t_{max})/M_{G0}$ of the maximum increases as Δ is made smaller and also decreases with ω.

These features of $M_G(t)$ are of course not arbitrary, because M_G is the result of a physical model, the linear model with infall. To understand this, therefore, one must observe the nature of the corresponding infall rates $f(t)$. From equation (15) it is evident that $f(t)$ may also have a maximum at $t > 0$. If so, it occurs at $t(\max f) = (k - 1)/\omega - \Delta = t(\max M_G) - 1/\omega$. The maximum infall rate occurs earlier, by $1/\omega$, than the maximum in the gas mass. This may often be inapplicable, of course. In the case $k = 1$, the infall rate is exactly exponential

$$k = 1: \qquad f_1(t) = \frac{M_{G0}}{\Delta} e^{-\omega t} \tag{21}$$

so that M_{G0}/Δ becomes the parameterized initial infall rate. That is, the physical meaning of Δ for $k = 1$ is that it is the time required for the initial infall to increase the disk mass to twice its initial value. For larger values of k the initial infall $f_k(0) = k(M_{G0}/\Delta)$, so that the initial infall would double the disk mass in the time Δ/k. The infall will at first increase with time if $(k - 1)/\omega > \Delta$, assuming an exponential-like decrease at large time.

Because of the infall, the total mass $M_{tot} = M_G(t) + S(t)$ increases with time, reaching an asymptotical value as $t \to \infty$. For the case $k = 1$, equation (21) shows us that the total asymptotic infall is simply

$$\int_0^\infty f_1 dt = M_{G0}/\omega\Delta \quad ,$$

or a total asymptotic mass that is $1 + (\omega\Delta)^{-1}$ times the initial mass. For $k = 2$ the asymptotic value is $1 + 2(\Delta\omega)^{-1} + 2(\Delta\omega)^{-2}$ times the initial mass, with analogous results for larger k. In figure 6.1, the value $M_{tot}(t)/M_{G0}$ is also displayed as a function of time. The case $\Delta = 0.1$, $\omega = 0.4$ rises to 26 times greater than M_{G0}, which is the asymptotic result. The case $\Delta = 4$, $\omega = 0.3$ reaches the value 1.824 at $t = 15$, just shy of the asymptotic result 1.833. But the major distinction between these two cases is clear: in the first parameterization the infall greatly increases the initial mass, whereas in the second parameterization it is increased by only 82%.

One obvious attribute of these models is that infall is still occurring; and that is an astrophysical question that is not settled. However, the amount may be small. In figure 6.1, for example, the $\omega = 0.4$ infall at $t = 15$ is e^{-6} times the initial infall, corresponding to $M_{tot}(15)/f(15) = 10^{12}$ yr, or 0.1% per 10^9 yr. The present time scale is not much different for the other example in figure 6.1. Decreasing ω and/or increasing k increases the importance of infall today, however. It is also not clear physically how the infall should go at short times—does it increase before decreasing, etc.? But it is perhaps useful that for $k > 1$ the infall in this mathematical model can be made to have a maximum at $t > 0$. Larson's (1976) dynamic infall models showed, for example, a peak in the star formation rate (at $r = 7.5$ kpc) near 5×10^9 yr when about 20–50% of the final star mass was already accomplished (see also Tinsley and Larson 1978). Such features are easily parameterized by the standard mathematical model advanced here.

Before going on I would observe that it is possible to reduce this standard model to certain limiting cases that have provided the traditional analytic solutions.

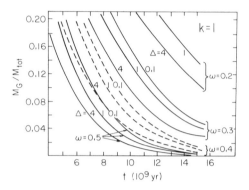

Fig. 6.2. Time dependence of M_G/M_{tot} for $k = 1$ models. Shown are three values $\Delta = 0.1, 1, 4$ belonging to separate choices for ω. Suitable choices of ω, Δ, T for our galaxy should result in 5% to 10% gas. Increasing ω and/or increasing Δ lowers the gas fraction at a given age. Although the pairs $\omega = 0.4$, $\Delta = 4$ and $\omega = 0.5$, $\Delta = 0.1$ are similar in this graph, they yield interestingly different results for other observables.

1. $k = 0$ corresponds to no infall, whereupon $M_G(t)$ is itself a simple exponential decrease. From (19) the stable concentration increases linear $(Z - Z_f) = y_z \omega t$. The radioactive chronologies also have simple solutions. Thoroughly studied, this model in its simplest form fails, primarily because $S(Z)$ has far too many low-Z stars (unless a high initial abundance Z_0 is postulated) and because astrophysical dynamics suggest that the disk mass should be continuously grown by infall.

2. $k \to \infty$ and $\Delta \to \infty$ such that $f/M_G = k/(t + \Delta) \to$ constant. Then $M_G =$ constant, making stars at the constant rate f. In this case (Larson 1972), $Z = y_z(1 - e^{-\nu}) \to y_z$ where $\nu = M_{tot}/M_G - 1$. This model is too simple in the opposite sense to the above: almost all stars have the same value of Z, namely, $Z = y_z$.

It is exactly to parameterize the large space between these two limits that an exactly soluble standard model is being advanced in the present work.

Figures 6.2 and 6.3 are designed to help quantify the parameter space of this model. Figure 6.2 shows, for the specific case $k = 1$, how the gas fraction $M_G(t)/M_{tot}$ declines with time for three different values of $\Delta = 0.1, 1, 4$ belonging to four different values of $\omega = 0.2, 0.3, 0.4, 0.5$. I have shown the large t and small gas fraction portion of these curves. They are mostly unremarkable since the trends are so evident. For the case $k = 2$, a very similar

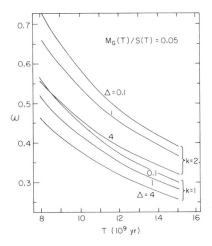

Fig. 6.3. The locus of models (ω versus T) resulting in a gas/star ratio $M_G(T)/S(T) = 0.05$. Here T is the galactic age rather than time. Loci for three values of $\Delta = 0.1$, 1, 4 are shown for each of two values of $k = 1$, 2. Such a relation allows a quick choice of parameters suitable to our galaxy. For example, the $k = 1$, $\Delta = 1$ locus is fitted approximately by $\omega \approx 4.3/T$, whereas the ωT products for the other $k = 1$ curves are 4.5 ($\Delta = 0.1$) and 3.9 ($\Delta = 4$). To obtain instead a gas/star ratio of 10%, the sets of curves should be lowered so that (approximately) the new $\Delta = 0.1$ curve for each k falls near the displayed $\Delta = 4$ curve for that k. More exactly, the approximately constant ωT product to achieve 10% gas falls to 3.7 for the $k = 1$, $\Delta = 0.1$ curve. For $k = 2$, 5% gas is achieved at $\omega T = 5.9$ ($\Delta = 0.1$), 5.5 ($\Delta = 1$), and 4.8 ($\Delta = 4$).

suite of curves obtains, except that (approximately) the labels on ω must be increased by about 0.06.

Figure 6.3 displays a related consideration. Suppose one requires instead a set of parameters that make today's gas fraction about 5%. This figure shows the locus of ω versus galactic age T for which the gas mass $M_G(T)$ is exactly 5% of the star mass $S(T)$. It does, in that sense, delineate the parameter space (for the case $k = 1$ and $k = 2$) appropriate to our own galaxy. The case $k = 3$ would demand still larger values of ω for each of the three choices $\Delta = 0.1$, 1, 4. And I also note, although it is not shown, that the loci corresponding to 10% gas would be found at lower ω in approximately this way: lower the three curves $\Delta = 0.1$, 1, 4 until the new curve for $\Delta = 0.1$ falls where the old curve for $\Delta = 4$ is. Then the gas mass is about 10% at T. Understanding this, one can almost by inspection select from figure 6.3 a model for our galaxy having ages between 8 and 15 Gyr and gas fractions between 5 and 10%. Exact parameters if needed are easily calculated from the elementary solutions given in equations (14) and (16).

When a reliable theory of galactic accretion has been established, it will be instructive to attempt to fit that $f(t)$ with appropriate choices of k, Δ, and ω. Until then we can only hope that other observational constraints can narrow the description. It is in fact with that intent that I have sought an exactly soluble parameterization in which the nuclear cosmochronologies can also be evaluated.

6.3.2 Age-Metallicity Relation

The metallicity $Z(t)$ of newly forming stars increases with time according to equation (19) for these models. One test of chemical evolution is the metallicity of stars according to their age. Although the ages of stars are not easy to obtain with assurance, the problem has been studied many times. In figure 6.4 the $[Fe/H] \equiv \log (Fe/H)/(Fe/H)_\odot$ is displayed against age using Twarog's (1980) data. For comparison I have plotted equation (19) setting $Z_f = 0$ on the right-hand ordinate, where $[Z/Z_{15}] \equiv \log (Z/Z_{15})$ and $Z_{15} = Z(15)$ is the metallicity at $T = 15$, when the gas has fallen to 5% in the model $k = 1$, $\omega = 0.3$, $\Delta = 1$. I have arbitrarily set $[Z/Z_{15}] = 0$ at $[Fe/H] = 0.1$. Although the resulting fit is very good, it is not strong evidence for the specific parameters $k = 1$, $\omega = 0.3$, and $\Delta = 1$, because many other sets of parameters also give good fits. Nonetheless one sees that this model is suitable for describing these data and for eliminating that portion of parameter space that does not fit it.

6.3.3 $S(Z)$ versus Z

Because $Z(t)$ grows quasilinearly with time, one might fear the same excess of low-Z dwarfs that plagues the simple closed $k = 0$ model, where $Z(t)$ is linear in time. But that problem does not occur because the early rate of star formation can be small in these infall models. Star formation at first increases with time in those models for which $M_G(t)$ has a maximum at $t > 0$, such as the $\Delta = 0.1$ case in figure 6.1. The way to make this test exactly is, of course, to consider equations (16) and (19) as parametric expressions for the normalized function $S(< Z)$ as a function of $\log Z$. This function is displayed in figure 6.5 for three different values of k. Case $k = 0$ is the simple closed exponential model for which too many low-Z stars result (assuming $Z = 0$

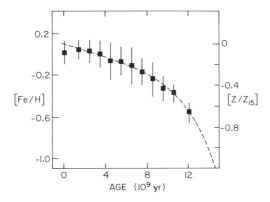

Fig. 6.4. The age-metallicity relation for stars in the galaxy taken from Twarog (1980) is displayed as points with error bars. The dashed curve is $\log(Z/Z_{15})$, the calculated $Z(t)$ normalized to its value at $t = 15$, when $[\text{Fe/H}] = 0.1$, for the model $k = 1$, $\omega = 0.3$, and $\Delta = 1$. The absence of lower observed metallicity can be attributed, in this model, to the dearth of low-Z stars.

initially), and is independent of ω and the parameter Δ (which does not exist for $k = 0$). These three k values are compared for the case $Z_f = 0$, which is the hardest case to make right. The curve $k = 1$ is calculated for the specific choice $\Delta = 1$, $\omega = 0.3$ in terms of $\log Z/Z_{15}$, where Z_{15} is the concentration at $t = 15$, when the gas mass is 4%. The fraction of stars with $S(< Z)$ is very much smaller, satisfactorily smaller, for small metallicity. The case $k = 2$, calculated with $\omega = 0.4$ and $\Delta = 1$ to obtain the same final mass fraction in gas, is even more deficient in low-Z stars. Thus, there is no "G-dwarf problem" in this model. That infall is capable of solving that problem has been known since Larson's (1972) original argument. What is being put forward here is an exactly soluble realistic model for evaluating the data. The data themselves, as far as I can tell in a preliminary study, fall intermediate to the $k = 1$ and $k = 2$ cases in figure 6.5. Of course, if $Z_f > 0$, the curves are compressed toward higher Z/Z_{15}, also alleviating the problem.

There is one very interesting feature of figure 6.5 that must be remarked upon. The $k = 1$, $\Delta = 1$ curve applies not only to $\omega = 0.3$, for which it was computed, but also for other values of ω provided that Z is normalized not to Z_{15} but to the Z obtaining at the same final gas fraction. Let me be specific. The case $\omega = 0.4$ instead of $\omega = 0.3$ reaches the 4% gas fraction at $t = 9$ instead of $t = 15$; and the curve $S(< Z)$ as a function of Z/Z_9 is essentially

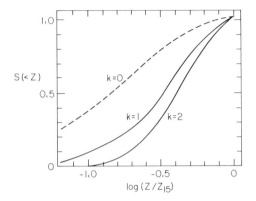

Fig. 6.5. The fraction $S(< Z)$ of stars that formed with metallicity less than Z displayed as a function of log (Z/Z_{15}). The linear model without infall ($k = 0$) has far too many stars having $[Z] < -0.7$ to fit the observations, but the $k = 1$ and $k = 2$ models chosen to achieve 5% gas mass at $t = 15$ do very well. The data would seem to lie roughly between the curves $k = 1$ and $k = 2$. Nonzero values for the initial abundance ($Z_0 \neq 0$) or infall abundance ($Z_f \neq 0$) would make each curve fall even more steeply toward low metallicity.

identical to the one shown as a function of Z/Z_{15}. The identity is not exact, because there exist two scales of time, Δ and ω; but if both are scaled appropriately, the time can be made to vanish explicitly from the $S(< Z)$ relation. The elimination of time for this particular $S(Z)$ relation has been known for a long time and was studied in detail by Lynden-Bell (1975).

6.3.4 Age Distribution and Mean Age of Dwarfs

In a linear model, where the birth rate of dwarfs is proportional to $M_G(t)$, the age distribution of dwarfs has the same functional shape as M_G. Simple closed models therefore have most of the dwarfs born at the beginning, because M_G decreases monotonically in time in those models. The present mathematical model has more realistic properties, since $M_G(t)$ may increase to its maximum at $t(\max M_G) = k/\omega - \Delta$, such as in figure 6.1. It need not be displayed further. On the other hand, the age distribution is a notoriously difficult observational problem.

It is interesting that the mean age of the dwarfs is also easily evaluated in this model. The mean formation time is

$$<t> = \frac{\int_0^t t\, dS/dt\, dt}{S(t)} = \frac{\int_0^t tM_G dt}{\int_0^t M_G dt} = \Delta \frac{I_{k+1}(t,-\omega) - I_k(t,-\omega)}{I_k(t,-\omega)} \qquad (22)$$

whereas the mean age $<\text{age}> = t - <t>$. Again, one needs only evaluate the simple functions I_k stated earlier for parameter $\lambda = -\omega$. Two numerical examples will suffice. Taking for the sake of discussion an age $t = 15$ for the galactic disk, the following mean stellar ages result from equation (22):

$$k = 1: \omega = 0.3,\ \Delta = 0.1 \qquad <\text{age}> = 9.20\ ,$$

$$k = 2: \omega = 0.3,\ \Delta = 0.1 \qquad <\text{age}> = 7.07\ .$$

Demarque and McClure (1977) found a mean age of 5 ± 1 Gyr for the disk galaxy NGC 188. Whether this be relevant to our galaxy or not, it is evident that the average dwarf age may be much less than the galactic age (in this example taken as 15). In fact in the $k = 2$ example the average age 7.07 is even less than it would have been for a constant star formation rate, showing that in that case more than half of the stars were born in the second half of the galactic age.

A related concept is the ratio of the average prior rate of star formation to the present rate:

$$\frac{<\psi(0-t)>}{\psi(t)} = \frac{S(t)/t}{\omega M_G(t)} = t^{-1}\left(\frac{\Delta}{t+\Delta}\right)^k e^{\omega t} I_k(t,-\omega)\ , \qquad (23)$$

where the constant factor $(1 - R)$ has been divided from numerator and denominator in the first equality and the second equality comes from the ratio of (16) to (14). For example, in the model $k = 1$, $\omega = 0.3$, and $\Delta = 0.1$ the ratio at $t = 15$ is $<\psi>/\psi = 4.3$, the ratio S/M_G being 19.25 at $t = 15$ in that model.

6.3.5 Metallicity Gradients

I have presented this "standard model" as if the parameters k, Δ, ω are fixed for the entire galaxy. That is likely to not be reasonable (e.g., Tinsley and Larson 1978). In that case we may choose them for studies of the solar neighborhood to be the values that were appropriate to infall and star formation in our annular zone. During considerations of an entire galaxy, however, whose gradients in both metallicity and gas fraction are known to exist, we need only choose the parameters to vary over the disk in a sensible way. One problem with this is that the parameter k is not continuous unless numerical complications in the evaluation of I_k are acceptable. My point is only to state that the abundance gradients can be modeled parametrically, although I will not do so here.

One might at first argue that imagining the parameter ω to be a function $\omega(r)$ of radius physically violates the simple assumption of the linear model, that $\psi(t) \sim \omega M_G(t)$. That is not necessarily the case, because the scale height of the disk also varies with r. And therefore, the mass density varies with r even if the surface density were constant. For the linear model to be physically meaningful it is sufficient that at fixed r, where the scale height $h(r)$ might be considered fixed, the star formation rate be proportional to $M_G(t)$ at that value of radius. The increase of gas fraction with radius ensures the existence of a metallicity gradient. Tinsley and Larson's (1978) discussion of their numerical models is very relevant and interesting for the question of trying to establish an analytic standard model.

6.3.6 Nuclear Cosmochronology

A major asset of this mathematical model is that the cosmochronologies can be evaluated, at least in the case of metal-poor infall. Although equation (18) is a nice solution for the concentration of a radioactive species, it is not physically plausible to assume that Z_f is constant for a radioactive species in the infall—unless, of course, $Z_f = 0$. So in this work I restrict myself to infall of metal-poor gas only, such that $Z_f \ll Z$, in which case $Z_f = 0$ may not be unphysical. This is a problem on which more knowledge is needed. How much nucleosynthesis, metallicity, etc., resided in the halo gas that later accreted to the disk? In any case, for numerical example take equation (18) with $Z_f = 0$. Likewise take $Z_0 = 0$ (no initial spike).

The real importance of nuclear cosmochronology is that time is an essential ingredient. Others have shown that for a wide class of models time can be eliminated in considering the interrelations between metallicity, star mass, gas mass, the distribution $S(Z)$, etc. But not so nuclear cosmochronology. It is for that very reason that I have sought a realistic standard model in which their concentrations can be evaluated explicitly. As far as I know, only the simple closed exponential model for the gas and the model of constant gas and star formation fed with constant infall have been solved exactly. These new models span the space in between in a specific useful parametric representation. To avoid from the outset a recurring confusion, one must distinguish between the "exponential models of cosmochronology" popularized by Fowler (1972) and the exponential gas of the closed linear model. In the latter $k = 0$ model the concentration $Z(t)$ is linear in t, and the radioactive concentrations also have relations that Fowler (1972) would term "constant nucleosynthesis." Fowler's point was quite different. Stated very simply it was that if the age distribution of the elements *in the solar system* (not in the galaxy as a whole) is monotonic, why not approximate that monotonic function by an exponential with a suitably chosen parameter. By doing so Fowler delineated much of the range of possibilities in nuclear cosmochronology. But that parameterization is not so easily and directly tied to other aspects of galactic evolution as is the model I advance here, where everything is physically motivated and exactly soluble.

The essence of nuclear cosmochronology is not simply the gas concentrations $Z(t)$, for they depend upon nuclear yields y_z that are difficult to calculate. It is rather in the ratio of the concentration $Z(t)$ to the concentration it would have had if the nucleus were stable that the time information lies. I therefore define the "remainder"

$$r(t) = \frac{Z(t, \lambda)}{Z(t, \lambda = 0)} = \frac{k+1}{\Delta} \frac{e^{-\lambda t}[\Delta/(t + \Delta)]^k I_k(t, \lambda)}{(t + \Delta)/\Delta - [(t + \Delta)/\Delta]^{-k}} , \qquad (24)$$

where the second equality comes from the ratio of (18) to (19) with infall Z_f set to zero. This remainder r may as a mnemonic be thought of as "the fraction remaining," with the understanding that it refers to the gas at that time, not to the total galactic content.

There is a significant and fascinating feature to equation (24); namely, it does not depend upon ω. This result is known historically only for the simple $k = 0$ model without infall, wherein $r_0 = (1 - e^{-\lambda t})/\lambda t$ independent of the exponential rate of decrease of the gas (true in that model). But we here see that the same result is true for this entire class of linear models provided that the infall is metal-free ($Z_f = 0$) and the initial gas is metal-free ($Z_0 = 0$). This rather surprising result is best seen from the general solution, equation (11), which shows that, if $Z_0 = Z_f = 0$, the ratio $r \equiv Z(t, \lambda)/Z(t, 0)$ is independent of the rate ω for the gas mass to convert to stars. As a result one finds that the nuclear cosmochronologies depend only upon the parameters k and Δ in this class of standard models. Thus, a correct

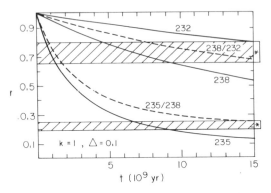

Fig. 6.6. The remainder $r(t) \equiv Z(t, \lambda)/Z(t, 0)$ compares the radioactive abundance in the gas to the value it would have had were it stable. The time dependence in the model $k = 1$, $\omega = 0.3$, $\Delta = 0.1$ is illustrated for ^{235,238}U and ^{232}Th. Galactic cosmochronology depends strongly on the ratios of remainders $r(235)/r(238)$ and $r(238)/r(232)$, which are shown as dashed curves. Horizontal bands marked by boxes on the right-hand ordinate are the observed ratios of remainders *if* Fowler's (1972, 1978) production ratios are adopted. The specific model that is shown is therefore consistent with observations if the Sun were born at $t_\odot > 14$, which would require a 19 Gyr old galaxy. Model comparisons are in figure 6.7.

nucleocosmochronological fit determines constraints on k and Δ, whereas the other astronomical evidence constrains also ω.

Because of remaining uncertainties on the production ratios for the radioactive chronometers, I will not try to isolate a "best model" here, preferring instead to illustrate the behavior and mathematics. I will show two formats whose meanings are easily understood. In the first one I arbitrarily choose an interesting model and display the behavior of the various remainders as a function of time, illustrating for interest the time when the ratios fit those inferred for the early solar system. The decay rates per 10^9 yr of the species to be considered here are, in increasing order: $\lambda(^{232}\text{Th}) = 0.0499$, $\lambda(^{238}\text{U}) = 0.154$, $\lambda(^{235}\text{U}) = 0.972$, and $\lambda(^{244}\text{Pu}) = 8.48$. So in figure 6.6, I take as given the parameters $k = 1$ and $\Delta = 0.1$ and plot $r(t)$. For $\omega = 0.3$ these parameters yield 5% gas after $T = 15$, so it is an interesting model. The remainders are plotted for ^{232}Th, ^{238}U, and ^{235}U, as are the ratios $r(238)/r(232)$ and $r(235)/r(238)$. Each remainder decreases monotonically in time, with ^{235}U of course the fastest of the three. That behavior is expected and is familiar from the simple closed exponential gas models, for which the remainder is simply $r_0 = (1 - e^{-\lambda t})/\lambda t$. The several $r(t)$ do not decrease as rapidly as in the simple closed ($k = 0$) model, however. The remainders r_1 for $k = 1$ and r_0 for the simple closed $k = 0$ case both fall from unity at $t = 0$, with r_1 falling less rapidly, until by $t = 15$ the remainder r_1 is, for the case ^{235}U, a factor 1.85 times greater than the value r_0 for the closed model. In fact, although I do not do it here, this ratio would be the most convenient way to visually scale the results of the calculations: to display them as $r_k(t)/r_0(t)$, so that they are of order unity independent of the decay rate λ. This ratio is analogous in spirit to Fowler's (1972) exponential parameterization, wherein the deviations from r_0 are labeled by a second time scale. And I again note to dispel confusion that the value r_0 is sometimes called the value for "constant nucleosynthesis," not because galactic nucleosynthesis occurs at a constant rate, for it does not, but rather because the age spectrum of the elements in the gas is flat. But there is a big difference between these models and Fowler's parameterization—a difference in sign. Because my models have infall, the remainders r_k/r_0 are greater than unity; i.e., elements look "younger" than in the closed-model case. But in Fowler's parameterization the elements look "older" than the closed case because $r/r_0 < 1$ for those models. It is of interest to compare these remainders with estimates of them, and I will concentrate on Fowler's (1972) estimates because they are so well reasoned and have been so influential to the field of nuclear cosmochronology. The uranium isotope ratio in the gas at any time is

$$\frac{Z(^{235}\text{U})}{Z(^{238}\text{U})} = \frac{y(^{235}\text{U})}{y(^{238}\text{U})} \frac{r(^{235}\text{U})}{r(^{238}\text{U})} ,$$ (25)

a ratio that is known to have had the value 0.313 at 4.6 Gyr ago. Division of that ratio by the production ratio, which Fowler has long maintained should be $y(235)/y(238) = 1.42 \pm 0.19$, gives the required ratio of remainders when the solar system formed: $r_\odot(235)/r_\odot(238) = 0.22 \pm 0.3$. That band is illustrated in figure 6.6, where one sees that the gas ratio $r(235)/r(238)$ never falls to a value that low in that particular ($k = 1$, $\omega = 0.3$, $\Delta = 0.1$) model, although it just reaches the upper limit at $t = 15$. Similarly shown is the required value $r_\odot(238)/r_\odot(232) = 0.72 \pm 0.07$ obtained from an abundance ratio (now) Th/U $= 4.0 \pm 0.2$ and Fowler's revised (1978) production ratio $y(232)/y(238) = 1.8 \pm 0.1$, somewhat higher than his 1972 value 1.65 ± 0.15. The particular model displayed in figure 6.6 does pass into that band for $t > 8$ Gyr. However, the total model is virtually eliminated *if* one accepts Fowler's estimate of the uranium production ratios because the solar system would have to have formed at $t > 14$. But that illustrates how the model works.

A next objective would be to improve the sophistication by seeking a model that does work. That can be done by being arbitrary; but the parameters should not be free. It is the beauty of this model that, because everything can be calculated, other astrophysical constraints can be utilized more easily than in arbitrary parameterizations. To this end I will decide to restrict (ω, T, Δ) to sets of values that give 5% gas at $t = T$ (today), although 10% might be a better choice for the solar neighborhood. That can be done from the loci illustrated in figure 6.3 (which, of course, cannot be read sufficiently accurately from that figure but are calculated with a simple Newton-Raphson technique that runs essentially instantaneously on a typical desk microcomputer). Furthermore, if the galactic age is T, the ratios r are to be evaluated at $T - 4.6$ Gyr rather than at T. This is shown for three different choices of the pair (k, Δ) in figure 6.7. Remember that every point along each curve is a *different model*, because for each age T the value of ω is chosen (for that k and Δ) to give a gas mass of 5% at $t = T$. But because the cosmochronology is independent of ω, the values of ω for figure 6.7 are moot. Comparison of the curves shows that $k = 2$ produces higher r ratios than does $k = 1$ when restricted to the same Δ (0.1 in the case shown) and to $M_G(T)/S(T) = 0.05$. This is because M_G peaks later, at $t(\max M_G) = (k/\omega) - \Delta$, for $k = 2$, so that the elements are younger for $k = 2$. The other comparison in figure 6.7 shows that increasing Δ from 0.1 to 4 lowers the remainder ratios. What is even more evident is that none of these models result in an $r(235)/r(238)$ ratio at $T - 4.6$ that is in agreement with Fowler's estimate of the production ratio. Agreement is obtained only by making infall unimportant (either $k = 0$ or large Δ). This disagreement means either that the model solutions presented here are unrealistic in some respect or that the production ratio $y(235)/y(238)$ is actually smaller than 1.4 ± 0.2. That the latter may be the case was concluded on grounds related to beta delayed fission by Wene and Johansson (1976) who obtained $y(235)/y(238) = 0.89$, and then again by Thielemann, Metzinger, and Klapdor (1983) who obtain $y(235)/y(238) = 1.24$. Because my purpose here is to advance my convenient model, I will not enter into this nuclear question at this time; however, the reader may study Thielemann et al.'s (1983) thorough but readable discussion of this point. However, I must note that their calculated $^{232}\mathrm{Th}/^{238}\mathrm{U} = 1.39$

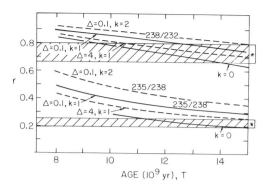

Fig. 6.7. The ratios of remainders at the time $T - 4.6$ when the Sun formed are shown for three different *families* of models as a function of their present age T. The solid curves, representing $k = 1$ and $\Delta = 0.1$ contain as the special case $T = 15$ the single model whose time dependence was shown in figure 6.6. The other models along the $k = 1$, $\Delta = 0.1$ solid curves are subject to the constraint that the pair of parameters ω, T be selected (as in figure 6.3) to yield 5% gas at $t = T$. The dashed curves illustrate changes in parameters k and Δ. Increasing k from 1 to 2 (at $\Delta = 0.1$) elevates the ratios of remainders, whereas increasing Δ from 0.1 to 4 (at $k = 1$) lowers the ratios. Models having metal-free infall (as shown) have difficulty matching the observed $r_\odot(235)/r_\odot(238) = 0.22 \pm 0.03$ unless the galaxy is older than 15 Gyr or the production ratio $y(235)/y(238) = 1.4 \pm 0.2$ (Fowler 1972, 1978) is too large.

is also much smaller than Fowler's 1.8, reducing the required value of $r(238)/r(232)$ to about 0.55 ± 0.05 and demanding an older galaxy.

A physical feature of the models that may be responsible for the discordance in $r(235)/r(238)$ is that the infall is nonzero, though decreasing, even at large time. But if infall ceases at some time, the later ratios of remainders will be smaller, as would be required for Fowler's production ratios. Such a procedure can also be carried out analytically by setting $f = 0$ at some time and joining the analytic solutions presented here to those elementary ones without infall. Results of this study will be reported in a second paper to be published elsewhere.

6.3.7 Extinct Radioactivities

Several important nuclei (i.e., ^{26}Al, ^{129}I, ^{244}Pu) are now extinct, and the issue has been often raised of their applicability to nuclear cosmochronology. In the present work I address only the issue of their expected *average* abundance in the interstellar gas. A last nucleosynthesis "spike" may be easily added to the models presented, but it is instead the mathematics of the continuous (average) nucleosynthesis that is our immediate concern.

The limit $Z(t)$ as $\lambda \to \infty$ (short-lived) is easily obtained, either from setting $dZ/dt = 0$ in equation (7) and following or by evaluating (11) for very large λ:

$$Z(t) \to \frac{y_z \omega}{\lambda + (f/M_G)} = \frac{y_z \omega}{\lambda + k/(t + \Delta)} \qquad \text{as } \lambda \to \infty \qquad (26)$$

for the models of this paper. The remainder for large λ becomes

$$r = \frac{Z(t,\lambda)}{Z(t,0)} \to \frac{k + 1}{(\lambda + [k/(t + \Delta)])[t + \Delta - \Delta^{k+1}/(t + \Delta)^k]}$$

$$\approx \frac{k + 1}{\lambda t} = \frac{(k + 1)\tau}{t} \quad , \qquad (27)$$

where the second approximation is true if $t \gg \Delta$ and $t \gg \tau \equiv 1/\lambda$. For the case $k = 0$ equation (27) goes to $r = \tau/T$, the oft-used estimate for the fraction remaining of a radioactive species. For $k > 0$, however, that estimate is too small by the factor $(k + 1)$ in these exact models. The reason for this lies in a smaller stable abundance rather than in the radioactivities themselves, for if we compare the remainders of two short-lived activities we obtain $r(\tau_1)/r(\tau_2) = \tau_1/\tau_2$. Thus, the factor $(k + 1)$ from equation (26) enters in the expected average ^{129}I/^{127}I ratio, but not in the ^{107}Pd/^{129}I ratio. This exact result becomes important for ^{129}I/^{127}I as k becomes a large integer.

Just how accurate are the approximations of equations (26) and (27)? They are exceedingly accurate for $\tau < 10^7$ yr. Perhaps the best indication is a direct comparison for ^{244}Pu, for which $\tau = 0.118$ Gyr. For example, the model $k = 1$, $\omega = 0.303$, $\Delta = 0.1$, and $T = 15$ gives in equation (24) at $t_\odot = 15 - 4.7 = 10.3$ the remainder $r_\odot(^{244}\text{Pu}) = 2.24 \times 10^{-2}$, which is to three significant figures the same result as the first equality in (27) and to two significant figures equal to the second approximation in (27), $r \approx 2$ $(0.118)/10.3 = 2.29 \times 10^{-2}$. So equation (27) is quite accurate even for ^{244}Pu.

The use of extinct radioactivities in cosmochronology is notoriously complicated by issues of the last "spike," if in fact it occurred at noticeable levels, and by the so-called "free-decay interval." These issues will not be addressed here except to recommend the mixing model of interstellar phases (Clayton 1983) as a better model of the free-decay interval, some model of which must be used to relate the average interstellar radioactive concentrations to those actually found in the forming solar system. On top of that, I think that an exact model from the family that I have here introduced, chosen to satisfy the various astronomical constraints, is a more realistic framework for nuclear cosmochronology than are more arbitrary parameterizations.

Returning to the ^{26}Al concentration in today's interstellar medium, I would add that the fraction remaining from supernova nucleosynthesis is changed only by the factor $(k + 1)$ from the simpler closed-model arguments given by Clayton (1984). My initial impression is that the range $k = 1$ to 3 will be adequate to find a good overall model, so that the factor $(k + 1)$ is nowhere near large enough to alter my previous conclusion that continuous galactic supernova nucleosynthesis is not the origin of the observed ^{26}Al gamma-ray line.

6.4 Conclusion

I have advanced an exactly soluble family of models for the chemical evolution of the galaxy; and I suggest adopting them as a "standard model" for evaluating the classic lines of evidence on the subject. They are almost as simple mathematically as the simple closed model so often used, but allow great freedom for parameterizing galactic infall and the history of the star formation rate. These models have the physical advantage of being solutions of the linear dependence of star formation rate upon gas mass, so that the exact parameterizations become a class of physical models in addition to a class of merely useful parameterizations.

In subsequent papers I plan to treat these and many additional results, subjecting them where possible to the known data. My primary concerns about this model are:

1) Perhaps infall does not continue today, but "turned off" at some time. These models are still exactly soluble, because the functional form $f/M_G = k/(t + \Delta)$ can be arbitrarily set to zero at some time, in which case an exact simple solution still exists.

2) Perhaps the infall abundance X_f is not constant in time, as it surely is not for radioactive nuclei, and as it is expected not to be if halo nucleosynthesis continued enriching the infalling gas for an appreciable time.

3) Perhaps the linear model is physically misleading and a higher dependence of star formation rate upon M_G is true.

4) Perhaps radial transport of matter in the disk is of such importance that the concept of a historically identifiable solar neighborhood is not meaningful.

5) How shall specific elements be treated when they do not arise equally from the same portions of the initial mass function, and is it itself constant in time?

Acknowledgments

This research was supported in part by NASA grant NSG-7361 and in part by the Robert A. Welch Foundation.

References

Clayton, D. D. 1983, *Ap. J.,* **268,** 381.

——. 1984, *Ap. J.,* **280,** 144.

Demarque, P., and McClure, R. D. 1977, in *The Evolution of Galaxies and Stellar Populations,* ed. B. M. Tinsley and R. B. Larson, p. 199 (New Haven: Yale University Observatory).

Fowler, W. A. 1972, in *Cosmology, Fusion, and Other Matters,* ed. F. Reines (Boulder: Colorado Associated University Press).

——. 1978, in *Cosmochemistry* (Houston: Robert A. Welch Foundation).

Larson, R. B. 1972, *Nature Phys. Sci.,* **236, 7.**

——. 1976, *M.N.R.A.S.,* **176,** 31.

Lynden-Bell, D. 1975, *Vistas in Astr.,* **19,** 299.

Mahoney, W. A.; Ling, J. C.; Jacobson, A. S.; and Lingenfelter, R. E. 1982, *Ap. J.,* **262, 742.**

Schmidt, M. 1959, *Ap. J.,* **129,** 243.

——. 1963, *Ap. J.,* **137, 758.**

Tinsley, B. A. 1975, *Ap. J.,* **198,** 145.

——. 1980, *Fundamentals of Cosmic Physics,* **5,** 287.

Tinsley, B. A., and Larson, R. B. 1978, *Ap. J.,* **221,** 554.

Thielemann, F.-K., Metzinger, J., and Klapdor, H. V. 1983, *Astr. Ap.,* **123,** 162.

Twarog, B. A. 1980, *Ap. J.,* **242, 242.**

Wene, C. O., and Johansson, S. A. E. 1976, *Proc. 3d International Conf. on Nuclei far from Stability,* Cargese (CERN-76-13), p. 584.

7

Cosmic Rays as a Possible Probe
of the Composition of Interstellar Matter

M. E. Wiedenbeck

Abstract

Recent advances in the measurement of cosmic ray isotopic abundance ratios which have significance for studies of stellar nucleosynthesis are reviewed. Emphasis is placed on isotopic anomalies reported in the elements neon, magnesium, and silicon. A variety of data is used to argue that the cosmic ray source composition may be a good tracer of the isotopic composition of present-day interstellar matter.

7.1 Introduction

Measurements of the relative abundances of the nuclides provide much of the observational basis for theories of stellar nucleosynthesis and of galactic chemical evolution. However, the compilations of "cosmic" or "universal" abundances (recent examples are found in Cameron 1982 and Anders and Ebihara 1982) which have been so influential in shaping these theories are based almost entirely on measurements of the composition of solid bodies in the solar system, that is, the Earth, the Moon, and meteorites. In using these data the tacit assumption is commonly made that they represent the average composition of matter in the interstellar medium.

The validity of this assumption is by no means obvious. The solar system condensed out of the interstellar gas approximately four and one-half billion years ago, and the composition of the matter which we now observe was "frozen-in" at that time. Furthermore, the matter which formed the solar system must have come from a rather small spatial volume in the Galaxy. Thus, unless the composition of matter were homogeneous throughout the Galaxy, one could not safely conclude that the abundances found in the solar

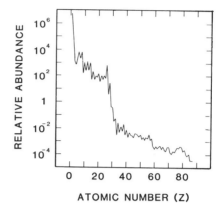

Fig. 7.1. Elemental composition of cosmic rays observed near Earth with energies ~1 GeV/amu. Cosmic ray abundances (both elemental and isotopic) can be influenced by acceleration and propagation effects, as well as by the composition of the source material. Thus careful selection of abundance ratios to be used for studies of nucleosynthetic effects is required.

system are characteristic of the average composition of the Galaxy at the time of solar system formation. In addition, the repeated processing of matter through stars and the ejection of the processed material back into the interstellar medium must alter the abundances of heavy nuclides in the Galaxy. Because the time from the formation of the solar system to the present is a significant fraction of the age of the Galaxy, it is possible that the composition observed in the solar system is also not the same as the composition of the local interstellar medium today.

Observations which can provide a direct measure of the composition of present-day interstellar matter are of great importance for checking the assumption that the solar system composition is "universal." Furthermore, if significant differences are found between solar system material and matter elsewhere in the Galaxy, the observation of galactic abundances will be essential for the further development of theories of stellar nucleosynthesis and galactic evolution.

An extensive body of information on the *elemental* composition of matter in a variety of galactic sites is available from astronomical observations in all bands of the electromagnetic spectrum and from measurements of the composition of galactic cosmic rays (figure 7.1). Elemental abundances alone are not, however, adequate for addressing many of the important questions of stellar nucleosynthesis. The different isotopes of an element are frequently synthesized under quite different physical conditions, and the ability to easily distinguish between different nucleosynthetic processes is largely lost if only the total abundance of the element can be determined. In addition, chemical fractionation affects all elemental composition investigations to some extent. (In fact, it is the same atomic properties that make elemental abundances easy to measure which also make chemical fractionation a serious problem.) The amount of fractionation can often be larger than the nucleosynthetic effects one would like to investigate.

In order to study nucleosynthetic processes it is important to use *isotopic* abundances since these are much more directly related to the physical processes to be studied. Furthermore, mass fractionation effects, if they occur at all, tend to be small and of a simple form for which corrections can easily be made. However, isotopic composition measurements tend to be difficult, and until very recently we had little information other than that obtained from the mass spectroscopy of bulk samples of solar system material.

In the past decade techniques for measuring isotopic abundances in galactic matter have been developed. I will discuss isotopic composition results which have been obtained from one such technique, the direct measurement of the isotopic composition of galactic matter arriving at Earth in the form of galactic cosmic rays. I will also discuss the possible significance of cosmic ray isotopic composition results for studies of stellar nucleosynthesis.

7.2 Cosmic Rays: Background Information

The modeling of the nucleosynthetic history of any sample of matter relies both on observations of the composition of that sample and on some knowledge of the post-nucleosynthetic processing of the material. The latter information is essential for distinguishing nucleosynthetic effects from any effects which subsequently alter the nucleosynthetic signature.

In the case of cosmic rays, the post-nucleosynthetic processing is known to be severe (see Simpson 1983 for a recent review). A small fraction of a thermal plasma is somehow selected for acceleration and then boosted to relativistic energies. These relativistic nuclei travel about the Galaxy under the confining influence of the interstellar magnetic fields for approximately 10^7 yrs. During this time they traverse significant amounts ($\sim$5–10 g cm^{-2}) of interstellar matter, in which they can undergo nuclear fragmentation reactions as a result of collisions with atoms of interstellar gas. In studying the nucleosynthetic processes which formed the cosmic ray source material, one is forced to consider the composition-altering effects of cosmic ray acceleration and transport. Our understanding of the injection and acceleration processes is still very limited; hence it is important to focus on the abundance ratios which are thought to be least sensitive to these effects. Abundance ratios between different isotopes of an individual element are thought to satisfy this criterion. Models of cosmic ray transport are largely empirical and do not address the basic physical processes in detail. They have, however, proven quite successful in accounting for the production of secondary cosmic rays during transport. However, since the observed abundances of many nuclides are dominated by secondary material, large uncertainties are introduced when one attempts to derive source abundances for these nuclides. Thus precise determination of source abundances is possible only for those nuclides whose flux arriving at Earth consists mostly of primary material.

Considerable progress has been made over the past decade in obtaining precise measurements of the isotopic composition of those cosmic ray elements which can provide the most information on stellar nucleosynthesis. At the same time new evidence has been accumulated concerning the post-nucleosynthetic processing of the matter which we see as cosmic rays. As a result of the combined progress in these two areas, cosmic ray composition studies are becoming a valuable new source of observational constraints for stellar nucleosynthesis theory. This new information is complementary to that obtained through the continuing study of the composition of solar system matter.

In § 7.3, I review the new developments in the study of the composition of cosmic ray material, and in § 7.4, I present some of the arguments which bear on our understanding of the post-nucleosynthetic history of this material. Finally, in § 7.5, I discuss the prospects for future cosmic ray observations which bear on questions of stellar nucleosynthesis.

7.3 Isotopic Composition of Cosmic Ray Source Material

The derivation of accurate cosmic ray source abundance ratios is subject to the two important limitations mentioned above. First, since chemical fractionation can significantly alter relative abundances of different elements, only the relative abundances of different isotopes of individual elements are likely

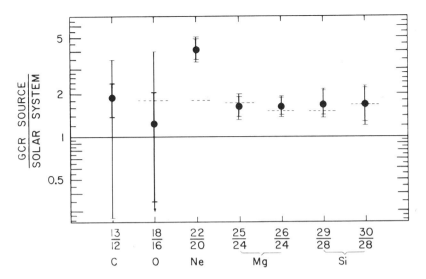

Fig. 7.2. Enhancement factors deduced for selected isotope ratios in the cosmic ray source (Wiedenbeck and Greiner 1981 a, b). Uncertainties are shown both with and without propagation uncertainties included. The indicated errors do not include the uncertainties in the solar system ratios used for normalization. Dashed lines indicate the predictions of one model for the synthesis of the cosmic ray material (Woosley and Weaver 1981).

to reliably reflect nucleosynthetic yields. Second, since nuclear fragmentation reactions during cosmic ray transport contribute "secondary" nuclei to the observed abundances, one can obtain accurate source abundances only for those nuclides for which the secondary contribution is relatively small. In the region of the periodic table accessible with present-day cosmic ray detectors, $1 \leq Z \leq 28$, at least five elements should be suitable for precise source isotopic composition studies. Three of these—neon, magnesium, and silicon—have been the most thoroughly investigated (Garcia-Munoz, Simpson, and Wefel 1979; Mewaldt et al. 1980a; Freier et al. 1980; Young et al. 1981; Wiedenbeck and Greiner 1981a, b; Webber 1982a). The other two, iron and nickel, require upcoming cosmic ray experiments to obtain the necessary statistical accuracy, although first results have already been obtained (Tarlé, Ahlen, and Cartwright 1979; Mewaldt et al. 1980b; Young et al. 1981; Webber 1981).

In reporting derived cosmic ray source abundance ratios, it is common practice to present these values relative to values of the corresponding ratios in solar system material, since the solar system composition is regarded as better understood and one expects that small differences between solar system and cosmic ray compositions might be accounted for by rather minor adjustments of present nucleosynthetic models. In keeping with this practice, I show in figure 7.2 the "enhancement factors" (defined as cosmic ray source ratio divided by solar system ratio) for the following isotopic abundance ratios: $^{13}C/^{12}C$, $^{18}O/^{16}O$, $^{22}Ne/^{20}Ne$, $^{25}Mg/^{24}Mg$, $^{26}Mg/^{24}Mg$, $^{29}Si/^{28}Si$, and $^{30}Si/^{28}Si$. There are two sets of error bars shown for each ratio. The inner set indicates the effect of cosmic ray measurement errors, while the outer set represents the combination of measurement errors with the uncertainty in correcting for the secondary contributions to the measured abundances. As anticipated, the uncertainties in the derived source abundances are large for the carbon and oxygen abundance ratios since the rare isotopes of these elements (^{13}C, ^{17}O, and ^{18}O) have large ($\gtrsim 70\%$) contributions from secondary cosmic rays. The ratios shown for the neon, magnesium, and silicon isotopes are quite accurate, however, since the observed abundances of all these isotopes (except ^{21}Ne, which we will not consider here) are mostly primary.

The five ratios shown for the neon, magnesium, and silicon isotopes all indicate a relative excess of the neutron-rich isotopes in the cosmic ray source. The $^{22}Ne/^{20}Ne$ ratio is particularly large, amounting to an excess by a factor of ~ 4 over this ratio in the neon-A component of solar system neon. The magnesium and silicon enhancements are smaller, each amounting to a factor ~ 1.5–1.8. It is particularly striking that the enhancement factors for these latter four ratios are all equal, to within the present uncertainties. Furthermore, it should be emphasized that the neon, magnesium, and silicon isotope ratios included in figure 7.2 are presently the only such ratios which have been determined in the cosmic ray source with an accuracy of better than about a factor of 2. Thus in every instance where it has been possible to look for differences from solar system composition with such an accuracy, significant differences have been found.

A number of efforts have been made to determine the source isotopic composition of cosmic ray iron both because iron is the next element in the periodic table which one expects to be amenable to precise isotopic composition determination, and because the synthesis of iron nuclei plays a key role in the evolution of massive stars. Thus far, the results have been limited by the resolution and/or the statistical accuracy of the observations. It has been clearly demonstrated (Mewaldt et al. 1980b) that ^{56}Fe is the dominant isotope

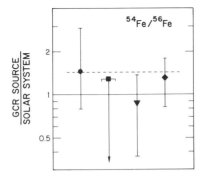

Fig. 7.3. Enhancement factors deduced for the ratio ^{54}Fe/^{56}Fe in the cosmic ray source from recent observations. The sources of the data are *circle,* Caltech *ISEE 3* experiment (Mewaldt et al. 1980*b*); *square,* Berkeley balloon experiment (Tarlé, Ahlen, and Cartwright 1979); *diamond,* Minnesota balloon experiment (Young et al. 1981); *triangle,* New Hampshire balloon experiment (Webber 1981). Errors are propagated observation uncertainties only. The dashed line shows the prediction of the Woosley and Weaver (1981) model with the same parameters as used for deriving the dashed lines shown in figure 7.2.

in cosmic ray iron, as it is in the solar system, so it is unlikely that much of the cosmic ray iron is produced under extreme conditions (such as in an environment of extremely high neutron excess).

The present status of investigations of the ^{54}Fe/^{56}Fe ratio in the cosmic ray source is indicated in figure 7.3. The deduced ratios are all consistent with the solar system value of the ^{54}Fe/^{56}Fe ratio, but the present uncertainties are too large to definitely exclude enhancements as large as those found for the magnesium and silicon isotope ratios. (Note that ^{54}Fe/^{56}Fe is thought to represent the ratio of a neutron-rich nuclide to an alpha-particle nuclide, as do the neon, magnesium, and silicon ratios presented above, since ^{56}Fe is thought to have been synthesized as ^{56}Ni, which subsequently decayed to produce the ^{56}Fe.)

The accuracy of source composition determinations for cosmic ray nickel isotopes is poorer than the accuracy of the ^{54}Fe/^{56}Fe ratio, due to the very limited statistical accuracy of the nickel observations (Ni/Fe $\approx$ 5%). At present it is not clear whether the cosmic ray nickel composition is consistent with the composition of solar system nickel (Tarlé, Ahlen, and Cartwright 1979; Young et al. 1981).

A number of models have been developed to explain the enhanced abundances of neutron-rich isotopes in the galactic cosmic ray source. I will not attempt to review the details of these models, other than to comment on a few of their general characteristics. It is evident that the builders of such models are sufficiently ingenious to explain the presently rather limited set of data on the isotopic composition of the cosmic ray source material with a variety of very different theories. These models are still of considerable importance to the cosmic ray experimenters when they are formulated with enough quantitative detail to predict which further observations would contribute most to distinguishing between the models and to constraining their parameters.

The models which have been proposed so far fall into three general categories: first, those which attribute the increased abundance of neutron-rich isotopes in the cosmic ray source to the evolutionary increase of these isotopes due to the repeated processing of galactic matter through massive stars (Arnett 1971; Woosley and Weaver 1981); second, those which explain the observed effects in terms of the production of cosmic rays by stars with characteristics differing from the average of those stars which synthesized the material from which the solar system formed (Cassé 1981 and references therein; Cassé and Paul 1982; Maeder 1983); and third, those which regard the galactic cosmic rays as a good sample of the interstellar medium and attribute their apparently anomalous composition to circumstances which caused the solar system composition to differ from that of the general interstellar medium (Olive and Schramm 1982).

Further progress in understanding the implications of cosmic ray composition for stellar nucleosynthesis can come from a variety of studies. Besides the obvious efforts to obtain experimental determinations of additional cosmic ray source abundances and quantitative theoretical predictions of the effects of various models, studies to determine what processing the cosmic ray material went through subsequent to nucleosynthesis are of particular importance. The types of post-nucleosynthetic processing assumed by the proposed models vary widely, and information constraining the possibilities can narrow the range of acceptable models.

7.4 Post-nucleosynthetic Processing of Cosmic Ray Material

A key question in attempting to choose among alternative models is whether there is a direct connection between the nucleosynthesis of cosmic ray matter and the acceleration of this matter to relativistic energies. For example, supernovae have long been considered a possible site for cosmic ray acceleration because the average power output of supernovae in the Galaxy is sufficient to maintain the observed cosmic ray energy density and because synchrotron emission from relativistic electrons in supernova remnants is observed. If cosmic ray nuclei were accelerated in supernova explosions, one would expect the cosmic ray composition to be strongly influenced by the composition of the pre-supernova star. Thus one might attempt to correlate cosmic ray composition anomalies with the abundance patterns peculiar to those stars which lead to most of the supernovae. Alternatively, if cosmic ray source material is not accelerated until long after it is synthesized and ejected from a star, one might expect the cosmic rays to more closely reflect the average composition of the interstellar medium.

Although the question of how closely cosmic ray acceleration is associated with supernovae is still the subject of much debate, there is now, in my opinion, a sizable collection of circumstantial evidence that cosmic ray acceleration is not directly connected with the stellar processes which synthesized the cosmic ray source matter and ejected it back into the interstellar medium. If these pieces of evidence are interpreted as showing that the cosmic ray source composition represents a sample of the general interstellar medium (at least over some limited volume of the Galaxy), then one can use the cosmic ray source isotopic composition values presented above as a direct measure of the isotopic composition of the elements neon, magnesium, and silicon in the present-day interstellar medium. If this interpretation is correct, cosmic ray isotopic composition observations should prove invaluable for testing models of stellar nucleosynthesis and galactic chemical evolution.

I will review a number of lines of evidence which point to the above interpretation of the nature of the cosmic ray source. It is the combined weight of these facts, rather than any single argument, which leads me to conclude that cosmic ray composition may be a good tracer of the composition of the interstellar medium.

7.4.1 Elemental Abundances and Their Correlation with Atomic Properties

Ratios between the abundances of elements in the galactic cosmic ray source and in the solar system vary over more than an order of magnitude, much more than the uncertainties in most of the abundance observations. It has been noted by a number of investigators (see Mewaldt 1981 and Simpson 1983 for reviews) that there exists a rather good correlation between these ratios and any of a number of related atomic properties such as first ionization potential, volatility, photoionization cross section, and so forth. This correlation is shown in figure 7.4 for the choice of first ionization potential as the organizing parameter. While the correlation is far from perfect, the general trend toward reduced abundances of elements with high first ionization potentials in the cosmic ray source is unmistakable. The physical mechanism responsible for producing this correlation is not presently known, but the mere fact that it involves the properties of the outer electronic shells of atoms shows that it must be a rather low temperature phenomenon ($\sim 10^4$–10^5 K).

The shells in a pre-supernova star in which most of the heavy elements reside are much too hot to provide these conditions. Thus one can argue that nucleosynthesis products must have been ejected from the star and allowed to cool substantially before cosmic ray acceleration took place.

It has been pointed out (Meyer 1983), however, that since refractory elements tend to form grains in a medium of the temperature we require, there could be a problem accounting for the normal abundances of such elements in cosmic rays. Meyer proposes that the mechanism which "injects" matter for cosmic ray acceleration is stellar flares and that the pool from which the matter is obtained lies in the photosphere of flare stars. For purposes of deciding how one should relate the composition of cosmic ray source matter to stellar nucleosynthesis processes, such a scenario may still be consistent with the hypothesis that the cosmic ray source reflects the general composition of the interstellar medium, since the outer atmospheres of many stars remain unaffected by the nuclear burning occurring deeper in those stars.

A related piece of evidence is obtained from observations of the elemental composition of the energetic particles emitted in solar flares. The ratios of the abundances seen in solar flares to the corresponding abundances found in the solar photosphere exhibit a correlation with first ionization potential which is quire similar to that found for galactic cosmic rays (Meyer 1981; Webber 1982b; Cook, Stone, and Vogt 1984). Thus the Sun seems to introduce a first

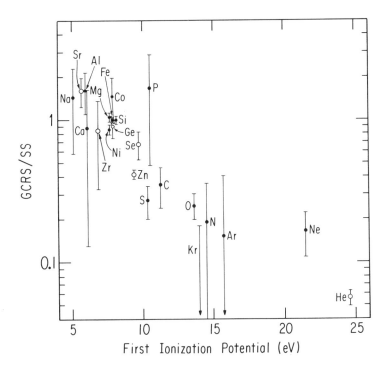

Fig. 7.4. Correlation of ratios of elemental abundances in the galactic cosmic ray source (GCRS) to those in the solar system with an atomic parameter (here chosen to be the first ionization potential) suggests that the GCRS material has spent time at low temperatures ($\sim 10^4$-10^5 K) after nucleosynthesis and before acceleration. Figure reproduced from Hinshaw and Wiedenbeck (1983), where details of the GCRS abundance calculations are given.

ionization potential bias on the elemental abundance pattern while causing (at least modest) particle acceleration, much as in Meyer's model. Furthermore, no clear indication of mass fractionation has yet been found in these processes[1] (Mewaldt, Spalding, and Stone 1984 and references therein; Simpson, Wefel, and Zamow 1983).

Another indication that nucleosynthesis products must be well mixed before they are accelerated to cosmic ray energies comes from the fact that elemental abundances in the cosmic ray source bear a great similarity to solar system abundances throughout the entire periodic table from carbon, nitrogen, and oxygen to the actinides, once the first ionization potential effects have been taken into account (Simpson 1983; Israel et al. 1983). Elements in different parts of the periodic table are synthesized under quite different stellar conditions, either in different regions of individual stars or even in different types of stars. It would be a remarkable coincidence for the products of such different nucleosynthesis processes as hydrostatic helium burning and r-process neutron capture to appear in essentially identical proportions in cosmic rays and in solar system material unless a very thorough mixing of stellar ejecta were to occur prior to cosmic ray acceleration. Such similar compositions are a natural consequence of mechanisms which accelerate cosmic rays from interstellar matter after the mixing of the ejecta from many stars has occurred.

7.4.2 Spectral Shapes

Observations of dominantly primary cosmic ray species (Garcia-Munoz et al. 1977; Engelmann et al. 1983) show that these elements have nearly identical energy spectra over almost two decades in energy.[2] These elements, including C, O, Ne, Mg, Si, S, Fe, and Ni, reside at widely varying depths in a pre-supernova star. The uniformity of the spectral shapes for these elements impose severe requirements on any model in which acceleration is presumed to take place in the supernova explosion itself. However, if there is time for the supernova ejecta to be well mixed before acceleration occurs, spectral uniformity is much more plausible.

Observations of ultra-heavy cosmic rays indicate that these elements have spectral shapes essentially the same as those of the lighter primary cosmic rays, although the data on the ultra-heavies are much more limited. In all likelihood, nuclides in widely separated regions of the periodic table are produced primarily by stars of different masses. This again suggests that it would be implausible that direct acceleration of cosmic rays at the stellar sites where they are synthesized would yield very similar spectra for such a wide range of elements.

7.4.3 The Time Delay since Nucleosynthesis

Several studies have been made of time delays since the synthesis of cosmic ray nuclei, using various radioactive nuclides as clocks (see review by Simpson 1983). One such approach relies on nuclides which can decay only by electron capture and which one believes should be produced in reasonable abundance in stellar nucleosynthesis.

If such nuclei are accelerated shortly (less than a half-life or so) after their synthesis, they will be stripped of their orbital electrons and become essentially stable. On the other hand, if acceleration is delayed and electrons remain available for capture by these nuclei, they will decay and will be depleted in the cosmic rays which are later accelerated. A study has been made of the elemental cobalt abundance using data from the *HEAO 3* satellite (Koch-Miramond 1981). It is expected that the pure electron capture isotopes ^{57}Co and ^{59}Ni will be produced during the nucleosynthesis of the iron peak. The ^{57}Co decays on a time scale of $\sim$1 yr, greatly reducing the total cobalt abundance. On a time scale of $\sim$10^5 yrs the decay of ^{59}Ni into ^{59}Co again increases the cobalt abundance. The *HEAO 3* group interpret the observed cobalt abundance as indicating a time delay of greater than $\sim$10^5 yrs, and argue that on this time scale the stellar ejecta will have been rather thoroughly mixed into the interstellar gas.

In another experiment on the *HEAO 3* satellite, the elemental composition of ultra-heavy cosmic rays ($Z > 28$) has been investigated. For purposes of determining how much time elapses between the synthesis and the observation of galactic cosmic rays, the actinide elements are particularly useful because all of the actinide isotopes are unstable. Furthermore, these nuclides can only be synthesized by rapid neutron capture. Studies of r-process nucleosynthesis indicate that the synthesized abundance ratio of actinide elements to lead-platinum peak elements should be significant ($\sim$5–20%). In the solar system a significantly smaller ratio is observed (less than 2%), presumably due to the billions of years that the actinides have had to decay. The *HEAO 3* cosmic ray results (Binns et al. 1982) indicate that the cosmic ray source composition is consistent with that found in the solar system, but is significantly less than expected from fresh r-process nucleosynthesis. Since alpha and beta decays of the actinides can occur during the $\sim$10^7 yrs between the acceleration of cosmic rays and the detection near Earth, the shorter lived elements (e.g., Np, Pu, Cm) are expected to have decayed, but the longer lived actinides should be present in the arriving cosmic rays if the material was freshly synthesized before acceleration. The observations suggest a long delay since the cosmic ray source material was synthesized, but more detailed analysis of the production and decays of the individual actinide isotopes is needed to make a more quantitative statement.

7.5 Future Prospects

The development of the experimental techniques needed to make precise measurements of the isotopic composition of galactic cosmic rays has expanded the scope of cosmic ray investigations. In the past, measurements of cosmic ray composition were used mainly as a probe of the astrophysics of cosmic ray origin, acceleration, and transport. Isotopic composition observations are now providing data which are, we believe, sufficiently insensitive to the processes of cosmic ray acceleration and transport that they can be used to study the nucleosynthesis of this material. Thus, instead of observing cosmic rays mainly for purposes of understanding cosmic ray phenomena themselves, some investigators are now utilizing the cosmic rays as a convenient vehicle for the transport of matter from otherwise inaccessible regions of the Galaxy to the neighborhood of the Earth where one can carry out detailed studies of the composition of that material in order to address questions of stellar nucleosynthesis and galactic evolution.

While it has, thus far, been possible only to study the source composition of a few elements, significant compositional "anomalies" have already been found. The next steps in exploiting this new experimental capability are clear. With detectors of the type which have already been flown it will be possible to investigate the isotopic composition of the key elements iron and nickel. Instruments capable of providing the necessary resolution and statistics to carry out these studies have already been constructed and are scheduled for launch in the second half of this decade. Proposals for still larger detectors, approaching 100 times the collecting power of instruments which have already been flown, are being made. Such an instrument would make possible studies of rare primary isotopes such as ^{57}Fe, ^{58}Fe, and ^{62}Ni and would permit exploratory investigations of the isotopic compositions of the lightest "ultra-heavy" elements, Cu and Zn.

The ultra-heavy cosmic rays ($Z > 28$) provide a large number of elements which can, with suitable instrumentation, be used for isotopic composition studies of cosmic ray source material. The criteria for obtaining data of use for nucleosynthesis studies (elements with two or more isotopes which consist mostly of primary material) are commonly satisfied because the ultra-heavy elements tend to have a sizable number of isotopes of comparable stability and because the abundance of these elements generally decreases with increasing atomic number, so there is little material to fragment and produce secondary cosmic rays. In addition, since ultra-heavy nuclides are mostly synthesized by neutron capture reactions, observations of these species will address a different aspect of stellar nucleosynthesis than the lighter species studies so far.

The experimental difficulties in measuring isotopic composition in the ultra-heavy region are, however, formidable. The absolute mass resolution of present-day instruments must be maintained while the fractional mass difference between adjacent isotopes is decreasing. Furthermore, this mass resolution must be achieved in an instrument with a collecting power at least 10^4 times that of instruments which have thus far been flown. The development of the necessary experimental techniques is being actively pursued by a number of groups, and the availability of long duration (several year) exposures of such large instruments in space should be practical before the end of this decade. Thus the prospects for cosmic ray investigations to further increase the observational base for stellar nucleosynthesis studies appear promising.

7.6 Conclusions

The potential importance of cosmic ray composition observations for studies of stellar nucleosynthesis has greatly increased over the past decade as the result of progress in two areas. First, isotopic composition measurements have begun to provide precise determinations of particular isotope abundance ratios which should be quite insensitive to post-nucleosynthetic processing. And second, a variety of measurements have provided new indications that the cosmic rays may well be providing a contemporary sample of matter from the interstellar medium. The combination of high precision isotopic measurements with a better defined framework in which to interpret them will, it is hoped, promote new interest in theoretical modeling of the nucleosynthetic history of the cosmic ray source material. There is great promise that such efforts will ultimately lead to an improved understanding of stellar nucleosynthesis and of the chemical evolution of our Galaxy.

Acknowledgment

This work was supported in part by NASA grant NAG5-308 and contract NAS5-20995.

Notes

1. The element helium is an exception: ^{3}He/^{4}He in ^{3}He-rich flares can be several orders of magnitude greater than its normal value.

2. Small differences in power law spectral indices have been reported in the high statistical accuracy study of Engelmann et al. (1983), but these may be within the systematic errors of their experiment.

References

Anders, E., and Ebihara, M. 1982, *Geochim. Cosmochim. Acta,* **46,** 2363.

Arnett, W. D. 1971, *Ap. J.,* **166,** 153.

Binns, W. R., Fickle, R. K., Garrard, T. L., Israel, M. H., Klarmann, J., Stone, E. C., and Waddington, C. J. 1982, *Ap. J. (Letters),* **261,** L117.

Cameron, A. G. W. 1982, in *Essays in Nuclear Astrophysics,* ed. C. A. Barnes, D. D. Clayton, and D. N. Schramm, pp. 23–43 (Cambridge: Cambridge University Press).

Cassé, M. 1981, *Proc. 17th Int. Cosmic Ray Conf.* (Paris), **13,** 111.

Cassé, M., and Paul, J. A. 1982, *Ap. J.,* **258,** 860.

Cook, W. R., Stone, E. C., and Vogt, R. E. 1984, to appear in *Ap. J.,* April 15.

Engelmann, J. J., Goret, P., Juliusson, E., Koch-Miramond, L., Masse, P., Soutoul, A., Byrnak, B., Lund, N., Peters, B., Rasmussen, I. L., Rotenberg, M., and Westergaard, N. J. 1983, *Proc. 18th Int. Cosmic Ray Conf.* (Bangalore), **2,** 17.

Freier, P. S., Young, J. S., and Waddington, C. J. 1980, *Ap. J. (Letters),* **240,** L53.

Garcia-Munoz, M., Mason, G. M., Simpson, J. A., and Wefel J. P. 1977, *Proc. 15th Int. Cosmic Ray Conf.* (Plovdiv), **1,** 230.

Garcia-Munoz, M., Simpson, J. A., and Wefel, J. P. 1979, *Ap. J. (Letters),* **232,** L95.

Hinshaw, G. F., and Wiedenbeck, M. E. 1983, *Proc. 18th Int. Cosmic Ray Conf.* (Bangalore), paper OG 5.2-7, to appear in Late Papers Volume.

Israel, M. H., Binns, W. R., Grossman, D. P., Klarmann, J., Margolis, S. H., Stone, E. C., Garrard, T. L., Krombel, K. E., Brewster, N. R., Fickle, R. K., and Waddington, C. J. 1983, *Proc. 18th Int. Cosmic Ray Conf.* (Bangalore), paper OG 6-39, to appear in Late Papers Volume.

Koch-Miramond, L. 1981, *Proc. 17th Int. Cosmic Ray Conf.* (Paris), **12,** 21.

Maeder, A. 1983, *Astr. Ap.*, **120,** 130.

Mewaldt, R. A. 1981, *Proc. 17th Int. Cosmic Ray Conf.* (Paris), **13,** 49.

Mewaldt, R. A., Spalding, J. D., and Stone, E. C. 1984, to appear in *Ap. J.*, May 15.

Mewaldt, R. A., Spalding, J. D., Stone, E. C., and Vogt, R. E. 1980*a*, *Ap. J. (Letters)*, **235,** L95.

——. 1980*b*, *Ap. J. (Letters)*, **236,** L121.

Meyer, J.-P. 1981, *Proc. 17th Int. Cosmic Ray Conf.* (Paris), **3,** 149.

——. 1983, CEN de Saclay preprint.

Olive, K. A., and Schramm, D. N. 1982, *Ap. J.*, **257,** 276.

Simpson, J. A. 1983, *Ann. Rev. Nucl. Part. Sci.*, **33,** 323.

Simpson, J. A., Wefel, J. P., and Zamow, R. 1983, *Proc. 18th Int. Cosmic Ray Conf.* (Bangalore), paper SP 2-6, to appear in Late Papers Volume.

Tarlé, G., Ahlen, S. P., and Cartwright, B. G. 1979, *Ap. J.*, **230,** 607.

Webber, W. R. 1981, *Proc. 17th Int. Cosmic Ray Conf.* (Paris), **2,** 80.

——. 1982*a*, *Ap. J.*, **252,** 386.

——. 1982*b*, *Ap. J.*, **255,** 329.

Wiedenbeck, M. E., and Greiner, D. E. 1981*a*, *Phys. Rev. Letters*, **46,** 682.

——. 1981*b*, *Ap. J. (Letters)*, **247,** L119.

Woosley, S. E., and Weaver, T. A. 1981, *Ap. J.*, **243,** 651.

Young, J. S., Freier, P. S., Waddington, C. J., Brewster, N. R., and Fickle, R. K. 1981, *Ap. J.*, **246,** 1014.

8

Nucleosynthetic Interpretations of Isotopic Anomalies

David N. Schramm

When one compares isotopic and elemental abundances obtained through different techniques in different classes of objects and events in the universe, the things one notes are the similarities and the differences. Since the material of which we have the greatest amount, and which has been subjected to the most detailed study, is solar system material, and of that solar system material the stuff we can get our hands on which has had the least chemical processing is the carbonaceous chondrites, we tend to refer to the relative heavy element and isotopic abundances in carbonaceous chondrites as our baseline. These abundances have historically formed the cornerstone of most cosmic abundance tables (see, e.g., Cameron 1983; Suess and Urey 1952).

Historically, observations disagreeing with this baseline have been referred to as "anomalies." For example, "anomalies" have been found in meteoritic inclusions (see the review by Lee 1979 and references therein) and in galactic cosmic rays (Wiedenbeck 1984) or the isotopic composition in the interstellar medium (Penzias 1980).

While chemical differences may be easy to explain due to chemical fractionation processes which do not require the high temperatures and densities necessary for nuclear processing, isotopic ones are significantly harder. Isotopic compositions have been obtained in meteorites both in the bulk composition and in primitive inclusions and have been determined for huge amounts of terrestrial material and in the solar wind, in moon rocks and dust, in cosmic rays, both galactic and solar, and in interstellar molecules, and most recently have been inferred in certain special cases of gamma-ray lines. They have also occasionally been inferred by molecular lines on cool stars.

Even though the solar system has provided the bulk of our data, we should not forget what our real scientific goal is, which is to understand the origin and chemical evolution of the universe, and this Galaxy in particular. To do this, perhaps the more important quantity is not the composition of the ubiquitous solar system material but the average composition of the interstellar gas and dust. This average is some sort of integral over stellar processes both from stellar mass loss and supernova ejecta and from novae, planetary nebulae, cosmic ray spallation, etc., all working to alter and supplement the initial light element abundances left over from the big bang. To understand variations relative to this interstellar gas we need to look at additional components of specific nuclear processing and whether it be via stellar processing or via irradiation, and also examine the possibility that there are even isotopic fractionations induced by some form of chemistry. It may be that variations in the interstellar gas and dust occur as a result of some regions having larger contributions from some type of objects in other regions and that the mixing has not been complete.

In trying to determine what the average is and what is anomalous, and in trying to understand how to produce the anomalies relative to the average, one has to beware of losing the forest in the trees. In particular, if one defines the bulk in such a way as to yield an observation to be anomalous, and the kind of processes that are necessary to produce that anomaly are extraordinarily peculiar, perhaps one needs to question the definition of the bulk. Remember also that hypothesizing that a particular anomaly is due to a very special fine-tuned process such as that anomaly coming from a particular zone of a particular type of star with nothing else added starts to sound contrived at best and at worst tends to imply that, when one integrates over a normal population of stars, if that process were sufficiently important to yield an observable effect, then maybe we should have redefined what our bulk composition was in the first place. Thus complementary to what one calls average interstellar and what one calls an anomaly are what one calls standard processes that produce the bulk composition and what one calls variants. It is imperative to put any such variants in the overall picture and not just look at some very special peculiar object. The bottom line in this sort of discussion is really the question, Is the solar system itself a variant, or is it representative of the average interstellar composition? This was a question that Keith Olive and I (1982) raised a couple of years ago, and it has still not been decisively answered.

The purpose of this paper is to reexamine this question. Let us first briefly review the standard model assumptions where it is assumed that the solar system abundances are the average. We will next look at the model where the solar system formed during the latter stages of a stellar association after one or more massive stars had exploded yielding a composition for the solar system more characteristic of the end of an OB association than of the average interstellar medium. We will then review selected anomalies and possible nucleosynthetic explanations for them, and we will conclude by showing that each model runs into problems that might be resolved with future observations.

8.1 The Standard Model

The standard model basically assumes that the abundances of the heavy elements are well represented for the most part by the carbonaceous chondrites when one normalizes these abundances through the use of the rare gases with the solar composition to obtain the overall average solar system abundances, which are then assumed to be the average galactic abundances. Thus one is saying that Cameron (1983) is approximately Anders (1982), which is approximately Suess and Urey (1956), which is approximately the average solar system composition, which is approximately equal to the average Galaxy composition. It is assumed that this composition has been obtained by adding to the hydrogen and helium, and traces of ^7Li coming out of the big bang, a well-mixed galactic soup with supernovae having added most of the heavies but with some admixture of debris from red giant winds, novae, planetary nebulae, etc. Anomalies occur when the anomalous object comes from a special source which has been well mixed so as not to show up in the overall average but was somehow retained, for example, by grains in the anomalous stuff.

8.2 The OB Association Evolution Model

In this alternative to the standard model it was noted that the Sun probably formed in some star association, perhaps an OB association, but in any case a stellar birth region which had massive stars. In stellar birth regions, star formation does not all occur instantaneously but in several stages, with at least some stars from the first stage having already evolved to the supernova stage while star formation is still going on in that region. This implies that the stars forming at the end of the association's lifetime will have significant contamination from the supernovae of the first stars forming in that association.

What has particular relevance to our discussion is that it might be argued that our Sun and solar system formed toward the latter part of such an association's history and thus had significant contamination of supernova debris. Thus our Sun's composition will be different from the average interstellar medium, which would be better represented by the first stars forming in the association. This is a model for the solar system which has been presented by Reeves (1978) and by Olive and Schramm (1982). The key aspect of this model is the question of how much material does the debris of the first few supernovae in the association mix. It might be noted here that the giant molecular clouds out of which stars form can be of the order of $10^5 \, M_\odot$. If the supernova debris mixes with the entire cloud (as was presumed in the initial calculations of Reeves), since the mass of heavy elements ejected in the supernova is less than $\sim 10 \, M_\odot$, the magnitude of any variation induced would be $\sim 10^{-4}$. However, the amount of material that goes into stars in one of these giant molecular clouds is only $\sim 10^3 \, M_\odot$. Thus, if the debris from the early supernovae mixes only with that material which goes into the stars, the magnitude of the heavy element enrichments is of the order of a few percent, which is comparable to the heavy element composition that the gas started with, and thus one has enrichment factors of order unity.

The reasons why Olive and Schramm argue for the latter case are multifold. First, they note that gas on the average in the Galaxy has been through a generation of stars only once. Thus it is to be expected that in order for the gas to reach its present heavy element abundance level of a few percent any generation of stars must be able to produce that amount of enrichment. It is also noted that calculations by Cox (1981) and others show that supernova debris will mix with the order of 1000 $M_\odot$ before it comes to rest. Thus, just from dynamical arguments and slowdown, one does not expect the supernova to freely mix with the entire $10^5 \, M_\odot$ but only with the material on the edge of the molecular cloud, the first 1000 solar masses. In fact, it is just this material on the edge which goes into making the stars of the association since star formation as observed seems not to be occurring throughout the entire cloud but primarily on the edge.

If we accept the conclusions of Olive and Schramm, the last stars to form in a star-forming region will have enrichments due to the first supernovae which go off prior to the stars' formation and the supernovae will increase these latter stars' heavy element abundances on the order of factors of 2 over the average of the interstellar medium. We then conclude that the stars in the Galaxy should probably have real variations of the order of factors of 2 in their heavy element composition. Whether such apparent observational variations are real is still problematic. It might, however, be noted that H II regions with average helium abundances, such as those associated with the new star-forming regions in the Orion nebulae, seem to have heavy element abundances that are approximately half the heavy element abundances of the Sun (Peimbert 1983). Since these new star-forming regions are presumably far younger than the 4.5 billion year old solar system and yet have lower heavy element abundances, it does appear that the solar system has been enriched in heavies relative to the interstellar gas. (However, one should be cautious here since the techniques for measuring heavy element abundances in H II regions and in the solar system are somewhat different. There is also the possibility that grains containing heavy elements in H II regions are not completely vaporized so that the gas composition of these regions might not be representative of their heavy element composition.)

Let us turn now to what other enrichments one might expect from supernovae. That is, what elements will the latter stars in the association have enriched? An entire generation of stars presumably produces the average interstellar medium, but if the last stars to form in the association have had extra contamination from supernovae which are the more massive stars in the association and have not had their complement of enrichment from the integral of the low mass stars of the association, then they will not receive a uniform enrichment in all heavy elements but only in those that are supernova produced. Such an effect is most likely to show up in those elements in which some isotopes are supernova produced and some are low mass star produced. Oxygen is an excellent example of such an element. ^{16}O is produced primarily by explosions of massive stars. ^{17}O and probably ^{18}O are more likely to be products of low mass stellar processes. Thus one might expect to have large enrichments of ^{16}O relative to ^{17}O and ^{18}O in supernova debris and thus in the last stars to form in a star-forming region. Carbon behaves similarly in that supernovae generate ^{12}C relative to ^{13}C. However, there is an additional complication with carbon since supernovae, at least for presently accepted values of the $^{12}C(\alpha\gamma)$ reaction rate, tend not to be able to produce enough ^{12}C; thus they seem to require an admixture of carbon production from intermediate mass stars. However, that does not alter the fact that the massive star explosions do not add ^{13}C but do add ^{12}C. In addition, chemically it might be noted that since massive stars add oxygen preferentially to carbon, one might expect to have a larger oxygen to carbon ratio in the ejecta of massive stars than in the average interstellar medium. Neon is another case in point, since ^{20}Ne is an α-particle nucleus and is a natural product of supernova explosions whereas ^{22}Ne is a product of secondary reactions such as helium burning on ^{14}N. Thus ^{22}Ne would be more likely to be found enriched in debris from low and intermediate mass stars and supernova debris would probably have higher ^{20}Ne to ^{22}Ne ratios than in the average interstellar medium. The bulk of the heavies, magnesium, silicon, aluminum, and sulfur, are thought to be produced only in supernovae and not in low mass stars, and thus their bulk isotopic composition is expected to be the same for both the stars with heavy supernova enrichment and the stars made out of the average interstellar medium. The iron abundance is not quite as clear because again massive stars or supernovae cannot produce enough iron and they may need to have an admixture coming from intermediate mass stars. The bulk of the s-process elements, with $A \lesssim 210$, similarly come from intermediate mass stars and not supernovae (Ulrich 1973; Truran 1982 and references therein). However, those s-process elements between iron and $A \sim 80$ are produced in type II supernovae (Couch and Arnett 1973; Wefel, Schramm, and Blake 1977).

The r-process nuclei are not quite as clear-cut because we don't really have a well-established astrophysical site for the r-process. It is clear that in supernova processing there are some zones where many neutrons will be released and thus significant neutron processing of the heavies will probably occur. Such processing produces significant amounts of many of the so-called r-process nuclei between the iron peak and $A \sim 210$. However, the actinide elements are not produced in such neutron processing, since the neutron fluxes are probably not strong enough to reach these levels (Blake et al. 1981). This point is also consistent with the fact that the ^{26}Al and ^{207}Pd producing events which added materials to the early solar system did not reset the ^{44}Pu clock. Thus, while we expect the massive stars to do some r-processing, they may not be the ones that reset the actinides. We do not really know where the

Table 8.1
Origin of Certain Isotopes and Elements

Low Mass Stars	High Mass Stars
$M \lesssim 10\ M_\odot$	$M \gtrsim 10\ M_\odot$
^{13}C	^{12}C
^{17}O, (^{18}O)	^{16}O
^{22}Ne	^{20}Ne
$O/C < 1$	$O/C > 1$
	Mg, Si, S, Ar, Ca, . . .
s-process $80 < A < 210$	s-process $A < 80$
? Fe ?	? Fe ?
	r-process $80 < A < 210$
? Actinide r-process ?	

actinide producing r-processing occurs. These results are summarized in table 8.1. Olive and Schramm (1982) went on to note that those elements most likely to be the ones affected significantly by the difference between high mass and low mass stars are also the ones which show the largest isotopic anomalies in various sources. Examples are the oxygen anomaly observed in meteorites (Clayton, Grossman, and Mayedo 1973) and the fact that the $^{12}C/^{13}C$ associated with s-process Xe may be different from the average $^{12}C/^{13}C$ in the solar system (Swart et al. 1983) and the $^{22}Ne/^{20}Ne$ anomaly observed in galactic cosmic rays. The direction of each of these observations goes in such a way as to be easier to understand if the solar system were enriched in massive star debris over the average interstellar medium. Let us now look at some special anomalies.

8.3 ^{26}Al

^{26}Al has a half-life τ_{26} of only 7×10^5 years. Thus the fact that Lee, Papanastassiou, and Wasserburg (1976) observed ^{26}Al to be present and alive when rocks form in the solar system implies that ^{26}Al had to have been synthesized within a few million years of the formation of the solar system. Theoretically there are at least two known classes of nucleosynthetic events that can produce ^{26}Al.

1. Explosive carbon burning in supernovae, which can produce $^{26}Al/^{27}Al \sim 10^{-3}$. This is also the site that is thought to produce the bulk of the ^{27}Al. It is associated with the massive stars that would go off at the beginning

of an OB association, and would thus be closely associated with star-forming regions.

2.   Novae. Novae can in special cases get ^{26}Al/^{27}Al ratios of the order of unity. However, novae are not the site of the production of the bulk of the ^{27}Al, so they are producing the ^{26}Al from preexisting aluminum and magnesium rather than building it up from light elements. Novae are associated with low mass stars and thus have no particular association with star-forming regions since by the time the stars in a star-forming region become novae the association has long since dispersed. Thus, if novae are the dominant origin of ^{26}Al, one would expect to have some sort of equilibrium ^{26}Al/^{27}Al ratio uniformly spread over the entire Galaxy and the observed Lee and Wasserburg value of ^{26}Al/^{27}Al $\sim$ 5 $\times$ 10^{-5} would then be implied to be characteristic of the entire Galaxy barring some fortuitous accident of a nearby nova happening to go off at the time of solar system formation.

It might be noted that with the newly measured low energy resonance production of ^{26}Al (Champagne, Howard, and Parker 1983), the nova production will be enhanced. In addition, "hot-bottom" red giants might even produce significant ^{26}Al which would be uniformly spread as in nova production.

Until recently the uniform abundance of ^{26}Al from novae appeared unlikely relative to the supernova model. However, a recent observation by Mahoney et al. (1982) using the *HEAO C* satellite implies that there may be a significant steady state abundance of ^{26}Al in the Galaxy. Clayton (1984) has argued that in fact the *HEAO C* observation is consistent with an abundance that's not too different from that observed by Lee and Wasserburg. Remember the steady state ^{26}Al is given by

$$\frac{^{26}\text{Al}}{^{27}\text{Al}} = \frac{P_{26}}{P_{27}} \cdot \frac{\tau_{26}}{T_{\text{Galaxy}}} \ .$$

This would require that the production ratio be of the order of unity and that the rate of nova explosions (or red giants) over the volume with which they mix be greater than the decay rate of ^{26}Al. In fact, this is not impossible since the number of novae in the Galaxy and the volume with which they mix could enable the mixing of the interstellar gas with the nova debris to go on on a time scale faster than the million year decay lifetime and it is trivially satisfied for red giants. Thus it may be that the ^{26}Al observation is telling us more about the synthesis in novae (or red giants) than in supernovae.

However, one still has to worry about the fact that novae or red giants do not produce the bulk of the aluminum and thus to have this happen means that most novae or many red giants must be producing ^{26}Al relative to ^{27}Al on the order of unity which was previously thought to require a nonstandard nova (see Truran 1978; Thielemann, Arnould, and Hillebrandt 1979). However, before totally accepting all that this implies, one should probably await the confirmation of the ^{26}Al observation by the Gamma Ray Observatory (GRO).

8.4 ^{107}Pd

Another key nuclear chronometer in the early solar system is ^{107}Pd which decays to ^{107}Ag in a few million years. A significant excess of ^{107}Ag observed in iron meteorites has been shown by Wasserburg and Chen (1981) to imply that ^{107}Pd was present when the solar system formed and went into iron regions, since paladium reacts chemically like iron and its decay produces Ag excesses in iron meteorites. Since iron meteorites are tens to hundreds of kilograms, it is clear that this enrichment of paladium is not due to retained interstellar grains. There is no present nova model which produces things as heavy as ^{107}Pd, in which case one seems to be drawn to a massive star origin for paladium. However, high neutron enrichments in novae and red giants and their subsequent effect on heavy element abundance have not been thoroughly explored, and given the new *HEAO C* results, it may be important to explore what novae might do to ^{107}Pd.

8.5 ^{16}O and FUN Anomalies

The two special cases we have discussed above are due to radioactive decay. In contrast, ^{16}O (Clayton, Mayeda, and Grossman 1974) and the FUN anomalies (see review by Lee 1979) are stable isotope anomalies. The retention of a stable isotope anomaly is far harder than the retention of a radioactive anomaly since radioactive anomalies will show up as long as the parent nucleus is still alive when final solidification occurs whereas to retain a stable isotope anomaly one has to limit the mixing, which is not limited in the case of radioactive anomalies (see discussion by Schramm 1978). Stable isotope anomalies not only are telling us that there are isotopic variations in the solar system, but also are telling us something about the mixing time scale relative to the condensation time scale in the early solar system. It might be noted that the ^{16}O anomaly which is the most prevalent of all isotopic anomalies in all meteorites and shows enrichments in ^{16}O of up to 5% is a natural product of supernovae, whereas ^{17}O and ^{18}O are not. Thus the ^{16}O might be understood in a straightforward way in the OB association model. The FUN anomalies are only noted in a few rare inclusions in carbonaceous chondrites, but when they are found it seems that almost every element in the inclusion is anomalous at $\sim 10^{-4}$. In some inclusions it appears that the neutron rich isotopes are enriched, and in others it appears that they are depleted. One solution may be that those which show enrichments such as inclusion EK1-4-1 have more of the supernova debris and those which show depletions in the neutron rich material such as C1 were condensations that were slightly more enriched in the material from the primitive cloud and should not have had as much admixture of supernova debris.

8.6 Cosmic Rays and ^{22}Ne

Cosmic rays are presumably also associated with supernova explosions, which are thought to be the accelerators for these particles. However, the material the supernova accelerates to make the cosmic rays is not the supernova debris itself but rather the interstellar gas directly surrounding the supernova. Since it is presumed that all supernova shock waves will produce cosmic rays via such mechanisms as that of Blandford and Ostriker (1978) or the modification of it due to Eichler (1981), the material getting accelerated is the material in the interstellar medium surrounding the typical supernova explosion. The very first stars to blow up in an OB association are the most massive stars, those perhaps on the order of 25–30 $M_\odot$. However, since stars all the way down to 8–10 $M_\odot$ blow up as type II supernovae, and perhaps even stars down to 6 $M_\odot$ blow up as some sort of supernovae, it is probable that the typical supernova blows up after the association has dispersed. Once a few supernovae go off in the association, they will disperse the cloud and prevent future star formation. Therefore the environment in which the typical supernova finds itself is really much more like the average interstellar medium, though slightly enriched because it is not too far removed from an association where a few massive stars have blown up, but it does not have the tremendous enrichments that might have been present when the last stars in the association actually formed. Thus it may be that the cosmic rays are a closer sample to the interstellar medium than the solar system and the solar system may show more significant supernova enrichment. This is contrary to the statements that were made several years ago (see, e.g., Hainebach and Schramm 1976; Arnett and Schramm 1973) prior to the Olive and Schramm OB model. If the cosmic rays are more like the average interstellar medium than the solar system, then one has a natural way of understanding the high ^{22}Ne/^{20}Ne observed in the cosmic rays by the Chicago and Caltech groups (see review by Simpson 1982). Since ^{22}Ne is a product of low mass and ^{20}Ne is a product of supernovae, it may have been that the solar system ^{20}Ne/^{22}Ne ratio has been heavily influenced by the excess ^{20}Ne coming from recent supernova explosions and the cosmic rays are more indicative of the average. One might also understand the high carbon to oxygen ratios in the cosmic rays since supernovae are more likely to make oxygen. Perhaps the solar system C/O ratio of the order of 1/2 is due to excess oxygen being added to the solar system, whereas the cosmic ray ratio of C/O ~ 1 is attributable to the average interstellar medium not having been as enriched in oxygen. This would be an alternative to using the ionization potential as an explanation of the C/O ratio in the cosmic rays.

8.7 ^{12}C/^{13}C Ratios in the Interstellar Medium

For many years radio astronomers have debated observed values of the ^{12}C/^{13}C ratio in the interstellar medium (see review by Penzias 1980). Different molecules and different clouds tend to give different values for this ratio. Although there have always been some astronomers who have favored interstellar values for this ratio that are similar to the solar system, there are also interstellar ratios reported which are significantly lower than those of the solar system. That is, many report significant enrichment in ^{13}C in the interstellar medium or, equivalently, that the solar system is enriched with ^{12}C. A few years ago the majority opinion seemed to favor a lower ^{12}C/^{13}C ratio in the interstellar medium of the solar system and it was argued by Dearborn, Tinsley, and Schramm (1978) that such ratios were not easily accommodated with standard galactic evolutionary models if the solar system was at the average 5 billion years ago, although nonstandard variations on these models could be made to work (see, e.g., Audouze 1977). As time has gone on, the average opinion on the average interstellar ^{12}C/^{13}C ratio has risen and may again be approaching that of the solar system although there are still observers who feel the average value is somewhat lower. It might be noted that the OB association model of Olive and Schramm arrives at an easy explanation for a low $^{12}C/^{13}C$ ratio in the interstellar medium by merely noting that the solar system could have been significantly enriched in ^{12}C.

It also might be noted that if the Olive and Schramm model is correct, then one would expect to have variations in the C/O ratio in different regions of the sky and different objects. Variations of the order of a factor of 2, both in the C/O ratio and in the $^{12}C/^{13}C$ ratio, might be expected. Thus the variations in different objects may be real rather than just statistical variations. Chemistry in clouds and condensation regions of the Galaxy that may have C/O $>$ 1 is quite different from chemistry in regions of the Galaxy where C/O $<$ 1 such as the solar system. This is because when C/O $>$ 1, once all the oxygen is taken up in CO, there is excess carbon left to make carbides and undergo reducing chemistry rather than oxidizing chemistry. There are even some meteoritic inclusions that appear to have been produced in regions where C/O might have been greater than 1 as discussed by Lattimer, Schramm, and Grossman (1977).

8.8 $^{12}C/^{13}C$ Ratios in s-Process Xenon

Swart et al. (1983) have noted that ^{13}C enrichments in meteorites seem to be correlated with s-process like xenon and krypton. Since s-process xenon presumably comes from low mass stars and similarly since ^{13}C presumably comes from low mass stars, this correlation is interesting. One argument says that these grains are from some particular low mass object which happened to make both s-process xenon and excess ^{13}C. The other option is to say that these are grains from the average interstellar medium, 4.5 billion years ago, prior to the addition of the massive supernova in the OB association. The first option of having it come from some special low mass star runs into a problem since no single zone in the star works and thus it requires careful mixing of two sites plus grain formation. While this is not impossible, it does require a carefully orchestrated model and one that is ubiquitous enough to produce examples in the solar system. The other option is of course consistent with the OB model of Olive and Schramm. These two points are discussed in detail by Dearborn and Schramm (1984).

8.9 Problems

Clearly there are several problems that remain to be resolved before we can know what is the best way of understanding the anomalies. Is the average solar system composition a good representation of the elements in the inter-stellar medium composition, or is the solar system significantly enriched in supernova debris, in which case some of the so-called anomalies may really be more representative of the average interstellar medium than the solar system is? One key aspect of this question will be the resolution of whether ^{26}Al is made in novae or red giants or in supernovae and whether there is a steady state abundance of ^{26}Al in the Galaxy. The resolution of this will probably have to await GRO. It might be noted here that both explanations, the low mass novae or red giants and the supernovae, yield live ^{26}Al when rocks form in the solar system, and thus both are consistent with the observation of Lee and Wasserburg. Neither requires presolar grains to be involved in the ^{26}Al explanation (Clayton 1984). Another problem that needs to be resolved is the fact that the ^{244}Pu time scale and thus the time scale for the production of actinides in the presolar r-process is a time scale of $\sim 10^8$ years between the last events that made the Pu and the formation of rocks in the solar system whereas the ^{107}Pd time scale is more like millions of years. Does this mean that the r-process for actinides is occurring at a separate site from the r-process for palladium? Or could it be that the ^{244}Pu is not made in special discrete events at 10^8 year intervals, with production ratios relative to the other actinides of the order of unity, but is instead also made in novae or low mass stars which are not so widely separated in time and no single source makes a large contribution? The possibility of low mass stars producing the actinide r-process seems to be what needs to be explored in far greater detail. Preliminary explorations of this in helium flashes have been done by Cameron and Cowan (1983). Such models for the r-process could significantly alter our time scale conclusions with regard to the origin of the r-process time scales in the early solar system. These time scales fit so well with the galactic rotation time of 10^8 years and the density wave theory of the Galaxy.

Another key problem that we clearly need to figure out is the cosmic rays. Are they more like the average interstellar medium or like supernovae? Further tests (Wiedenbeck 1984) on the isotopic composition of the cosmic rays and in particular on primary and secondary radioactive nuclei may eventually resolve this.

The final question is the question of whether the interstellar medium has evolved significantly in the last 4.5 billion years and whether that is an expla-nation for any differences which exist between the interstellar medium and the solar system. Or is it that the solar system was significantly supernova enriched and that there are many variations about the average in the present-day interstellar medium? Clearly more observations are critical here.

8.10 Conclusion

The study of anomalies and their origins is critical to our understanding of the evolution of the Galaxy, the origin of the solar system, the nucleosynthetic models of supernovae, novae, red giants, and other sources, and the origin of cosmic rays. At present we see that there are many alternative ways of looking at anomalies and we still do not really know what is anomalous and what is the bulk composition. But at least we are to the point where observational tests will eventually guide our selection as to what is an anomaly and what is not.

Acknowledgments

I wish to acknowledge my recent collaborators Keith Olive and David Dearborn as well as my former collaborators Kem Hainebach, Typhoon Lee, Ed Anders, Robert Clayton, Jim Lattimer, and Dave Arnett and of course my collaborators and mentors Willy Fowler and Jerry Wasserburg. Each of these people played a key role in shaping my views on the subject, although none is responsible for my utterances here. This work was supported in part by NASA grant NSG7212 and NSF grant AST-8116750 at the University of Chicago.

References

Anders, E., and Eberhart, P. 1982, *Geochim. Cosmochim. Acta*, **46**, 2363.

Arnett, W. D., and Schramm, D. N. 1973, *Ap. J. (Letters)*, **184**, L47.

Audouze, J., ed. 1977, *CNO Isotopes* (Dordrecht: Reidel).

Blake, J. B.; Woosley, S.; Weaver, T.; and Schramm, D. N. 1981, *Ap. J.*, **248**, 315.

Blandford, R. D., and Ostriker, J. P. 1978, *Ap. J.*, **221**, 464.

Cameron, A. G. W. 1982, in *Essays in Nuclear Astrophysics*, ed. C. A. Barnes, D. D. Clayton, and D. N. Schramm, p. 23 (Cambridge: Cambridge University Press).

Cameron, A. G. W., and Cowan, R. 1983, Center for Astrophysics preprint.

Champagne, A. E., Howard, A. J., and Parker, P. D. 1983, *Ap. J.*, **269**, 686.

Clayton, D. D. 1977, *Icarus*, **32**, 255.

——. 1984, this volume.

Clayton, R. N., Grossman, L., and Mayedo, T. K. 1973, *Science*, **182**, 485.

Couch, R., and Arnett, W. D. 1973, *Explosive Nucleosynthesis* (Austin: University of Texas Press).

Cox, D. P. 1981, *Ap. J.*, **245**, 534.

Dearborn, D. S. P. and Schramm, D. N. 1984, EFI preprint.

Dearborn, D. S. P., Tinsley, B., and Schramm, D. N. 1978, *Ap. J.*, **223**, 557.

Eichler, D. 1981, *Proceedings of the International Cosmic Ray Conference*, Paris.

Hainebach, K. L., and Schramm, D. N. 1976, *Ap. J. (Letters)*, **207**, L79.

Iben, I., Jr., and Truran, J. W. 1978, *Ap. J.*, **220**, 980.

Lattimer, J., Schramm, D. N., and Grossman, L. 1977, *Nature*, **269**, 116.

Lee, T. 1979, *Ref. Geophys. Space Phys.*, **17**, 1591.

Lee, T., Papanastassiou, D., and Wasserburg, G. 1976, *Ap. J. (Letters)*, **221**, L107.

Mahoney, W. A.; Ling, J. C.; Jacobson, A. S.; and Lingenfelter, R. E. 1982, *Ap. J.*, **262**, 742 (see also 1984, JPL preprint).

Norman, E. B., and Schramm, D. N. 1979, *Ap. J.*, **228**, 881.

Olive, K., and Schramm, D. N. 1982, *Ap. J.*, **257**, 276.

Peimbert, M. 1983, *Proceedings of the Munich Conference on He Abundances.*

Penzias, A. A. 1980, *Science*, **208**, 663.

Reeves, H. 1978, in *Protostars and Planets*, ed. T. Gehrels (Tucson: University of Arizona Press).

Schramm, D. N. 1978, in *Protostars and Planets*, ed. T. Gehrels (Tucson: University of Arizona Press).

Schramm, D. N., and Olive, K. 1980, *Proceedings of the Orion Symposium*, New York.

Seuss, H. E., and Urey, H. D. 1956, *Rev. Mod. Phys.*, **28**, 53.

Simpson, J. 1983, *Ann. Rev. Nuc. Part. Sci.*, **33**, 323.

Swart, P. K.; Grady, M. M.; Pillinger, C. T.; Lewis, R. S.; and Anders, A. 1983, *Science*, submitted.

Thielemann, F., Arnould, M., and Hillebrandt, W. 1979, *Astr. Ap.*, **74**, 175.

Truran, J. W. 1982, in *Essays in Nuclear Astrophysics*, ed. C. A. Barnes, D. D. Clayton, and D. N. Schramm (Cambridge: Cambridge University Press).

Ulrich, R. K. 1973, in *Explosive Nucleosynthesis*, ed. D. N. Schramm and W. D. Arnett (Austin: University of Texas Press).

Wasserburg, G., and Chen, T. 1981, *Proc. Meteoritical Society*, Bern, Switzerland.

Wefel, J. P., Schramm, D. N., and Blake, J. B. 1977, *Ap. Space Sci.*, **336**, 335.

Wiedenbeck, M. 1984, this volume.

9

Late Stages of the Evolution of Massive Stars: Challenges

W. David Arnett

Abstract

A number of questions are posed to focus attention on aspects of the evolution of massive stars which are not now well understood, but with effort might be.

9.1 Introduction

Although the theory of the advanced evolution of massive stars has advanced significantly (see Woosley and Weaver 1982; Weaver, Woosley, and Fuller 1983 for the state-of-the-art results), some serious problems still challenge our understanding. Rather than reiterate past successes in this area of research, this paper will attempt to begin a dialog concerning some of these problems.

The format will be a series of questions, each followed by some discussion, and in a few cases by what may be partial answers. The questions will be gathered into three sections, concerning observations, core collapse, and pre-supernova evolution. We focus on single star evolution, not because it alone is important, but merely because (1) it may well be a prerequisite for a deep understanding of multiple star evolution and (2) we find ourselves inundated by difficulties even at this level.

9.2 Some Observational Questions

A. *What is the yield of iron in a supernova explosion?* Theorists now routinely assume that the light curves of type I supernovae require about 0.5 to 1 $M_\odot$ of ^{56}Ni. Some type II supernova observations show an exponential decline at late times that is likely to be due to ^{56}Co decay. *Do the light curves of all type II supernovae which are observed to faint magnitudes show such an "exponential" tail?* These radioactive nuclei decay to ^{56}Fe, which should be little diluted in young supernova remnants. *Can we understand the physics of these young remnants well enough to place significant observational constraints on the iron yields?*

B. *Can gamma-ray lines be detected from supernovae and young supernova remnants?* The decay of ^{56}Ni leads to excited states of ^{56}Co which emit gamma-ray lines of characteristic energies. Other decays, such as ^{56}Co to ^{56}Fe, emit their own characteristic gamma-line spectra. If an explosion spreads the radioactive matter sufficiently fast, so that such gamma-rays escape without modification (for example, by Compton scattering), then such gamma spectra can be a direct probe of the relative abundances of the newly synthesized matter. While ^{56}Ni probably decays before transparency to gamma-rays, ^{56}Co is a promising source of transient gamma lines.

C. *What is the frequency of neutron star and black hole formation in stellar collapse events? What is the mass distribution of the condensed remnants?* We talk of neutron stars having masses near 1.4 $M_{\odot}$, the Chandrasekhar limit, and black holes having a mass $M \gtrsim 8\,M_{\odot}$ (from Cygnux X-1). The existence of a gap in remnant mass would have profound implications for understanding the collapse process. A "failed core bounce" (see Hillebrandt and Wolff, this volume) might swallow its mantle, or eject it by a "mantle explosion" or by some other process. The success of the core bounce mechanism seems finely tuned in mass, and might be expected to give a narrow range in neutron star mass. Well-determined and significantly different neutron star masses would have far-reaching implications.

D. *What is the gravitational wave signature of a supernova?* Presumably some supernovae leave no condensed remnant and have negligible gravitational radiation. In a collapse event, detection of the gravitational waveforms would allow us to probe the mass motion in the deepest interior. The extreme experimental difficulties are well known (see articles in Smarr 1979, for example).

E. *What is the neutrino-antineutrino signature of a supernova?* Again, the collapse events are the most promising. Because present-day detectors only probe a fraction of our own Galaxy, the expected event rate is of the order of one per century or worse. In an explosion there will be a burst associated with the emergence of the shock through the neutrino-sphere. Over a longer period (of order seconds) the neutron star will lose most of its lepton excess ("deleptonize"), so that the neutrino flux must exceed the antineutrino flux. Circulation currents could drastically modify this stage. It is not clear whether delayed accretion or mass loss occurs; this might significantly affect the observed spectrum.

F. *Can we identify the nucleosynthesis layers in supernovae and young supernova remnants?* Microscopic mixing to a uniform composition is a formidable task to accomplish on a supernova time scale. Observations are beginning to indicate that this fairly obvious theoretical inference may well be correct and that we may be able to identify various regions processed by different thermonuclear evolution. The nucleosynthesis processes reflect basic features of nuclear physics and are more fundamental than particular stellar evolutionary results. A search for examples of all the major stages (hydrogen, helium, carbon, neon, oxygen, and silicon burning) should be made. Then we should attempt to relate particular remnants to particular model sequences, a task which places heavy demands upon the abundance determinations throughout the objects.

9.3 Core Collapse

A. *How do the results of core collapse depend upon the characteristics of the initial models?* In attaining consistency between the various groups doing calcualtions of the collapse itself, it has been both necessary and desirable to use the same (or similar) initial models. There are indications that the collapse results are sensitive to the nature of these initial models (see Hillebrandt and Wolff, this volume), so that the range of possibilities should be defined and explored (see below).

B. *How do the results of core collapse depend upon inconsistency between the hydrostatic calculations which generate the initial models and the hydrodynamic calculations of subsequent evolution?* Representations of equations of state are never quite the same. Pressure deficits tend to reduce the effectiveness of the "core bounce" mechanism. A selective bias is possible here. If a hydrodynamicist finds that the initial model expands, some change is made so that it will not. If it collapses, the calculation continues. For example, Jin and Arnett (1985) argue that neglect of the Coulomb effect of the ions (in the hydrostatic calculations) can give a significant error of this sort.

C. *What happens to the shock when it becomes transparent to neutrinos?* To date there has been no detailed numerical calculation of the effect, on the shock, of the decoupling of neutrinos and matter, which must occur as it propagates to lower density. The diffusion approximation breaks down. The "neutrino-sphere" occurs at widely different radii for different neutrino energies rather than at a well-defined spherical surface. The neutrino emission becomes superthermal. Alpha-particles become an important constituent behind the shock. These complications make the problem formidable.

9.4 Pre-Supernova Evolution

A. *What are the products of stellar helium burning?* The abundances resulting from helium burning depend upon the ratio of $^{12}C(\alpha,\gamma)$ ^{16}O to $^4He(2\alpha,\gamma)$ ^{12}C. A higher $^{12}C(\alpha,\gamma)$ ^{16}O rate produces more ^{16}O, at the expense of ^{12}C. The ratio $^{12}C/^{16}O$ affects subsequent evolution; if there is little ^{12}C, then carbon burning (and neon burning) will be weak. The experimental situation is still difficult (see Fowler, and Rolfs, this volume); and the quoted errors allow significant variation in nucleosynthesis (Thielemann and Arnett, this volume).

B. *How do very massive stars evolve?* Stars of several hundred to several thousand solar masses we will call "very massive objects" (VMOs). Over the years, the belief in the existence of VMOs has fluctuated; at present they are of interest with regard both to presently observed luminous enigmatic objects (30 Doradus, η Carina, etc.) and to pregalactic and/or early galactic evolution (e.g., Carr, Bond, and Arnett 1984). Because of the dominance of radiation pressure, the effective adiabatic exponent γ approaches the value for marginal stability (4/3 for Newtonian gravity). The forces that restore the structure to its hydrostatic state become small. Even small forces, ignored for the evolution of lower mass stars, can become dominant for VMOs. Rotational forces, for example, might have important effects on the evolution. Mild instabilities might have dramatic consequences.

C. *What are the effects of "hidden hydrodynamics" on stellar evolutionary calculations?* In calculations of stellar evolution it is conventional to separate some hydrodynamic processes and model them in parameterized ways which suppress their hydrodynamic nature. Convection and mass loss are prominent examples. There are practical reasons for such procedures, but their ultimate justification must lie in an understanding of the hydrodynamic processes themselves, which presently eludes us.

To gain a new perspective, it is interesting to attempt to incorporate more of the physics of the hydrodynamic process directly into the evolutionary calculation. Traditionally convection is assumed to carry heat and mix abundances but not affect the "hydrostatic" structure. In the same spirit we can calculate a perturbative hydrodynamics (e.g., Arnett 1969) in which mass and energy are transported, but convection does not affect the unperturbed pressure balance equation. This means that the "ram pressure" of convection is ignored. To calculate over evolutionary time scales, the perturbative hydrodynamics must be done implicitly; this filters out sound and shock waves.

Can one calculate stellar evolution—with hydrodynamics—up to the photo-sphere? The answer seems to be yes for the very simple "perturbative hydrodynamics" just described. There are two approaches used in stellar evolution to treat the outer layers. One may ignore the difficulties by crude zoning near the surface. This does not allow one to construct an accurate H-R diagram, and for cool stars it can distort the evolution of the deeper regions by suppressing the outer convection zone. The traditional solution is to match the inner calculation to a hydrostatic envelope calculated separately. This renders awkward (impossible?) the treatment of pulsation and mass loss. Note that massive stars evolve to a stage where both of these phenomena occur.

Our experiment involved a 24 $M_{\odot}$ star (Population I), evolved to the red supergiant region. No external pressure was imposed. The radius was sensitive to variations in the opacity across the outer zone. The radiative flux at boundary K is, in standard notations,

$$F_K = -\left(\frac{ac}{3\rho\kappa}\right)_K \frac{\left(T_{K+1/2}^4 - T_{K-1/2}^4\right)}{\Delta r_K},$$

so that the opacity drop due to recombination ($\kappa \approx T^9$ or so) is reflected in a pronounced flux increase. Rather than approximate κ_K by a value at the center of the zone (with $\kappa(T_{K-1/2})$, for example), the surface temperature T_K was used.

The convective flux is

$$F = \rho v W$$

where v is the radial velocity of convective flow and $W = E + PV$ is the enthalpy. For high luminosity, cool stars (Hayashi, Hoshi, and Sugimoto 1962) the convective velocity approaches the sound speed $s \approx \gamma P / \rho$. In the structure equation, we should replace the pressure gradient ∇P by $\nabla(P + \rho v^2)$ to take account of the momentum transferred by convection. Now, if $v \approx s$,

$$\rho v^2 \approx \rho s^2 \approx \gamma P \approx P \ ,$$

so that neglect of this "ram pressure" is not justified. The Mach number v/s is largest at the top of the convection zone, so that dissipation should be large there too. One might imagine a churning, chaotic, dissipative region; with

$$v \approx s \approx 0.1 v_{escape}$$

the possibility of strongly driven mass loss is obvious.

Finally, pulsation and convection may not be separable processes for such a star. A further generalization of the "hydrodynamic" algorithm used here is needed to incorporate these neglected effects.

D. *Are there important "dynamic" effects of convection that are being missed?* It is presumed that, although convection is supposed to be a chaotic, hydrodynamic process, the longer term behavior is steady state (on average) and uniform on spherical shells (again, on average). The question above lumps the breakdown of these assumptions into the term "dynamic effects" which is a bit narrow.

There is a class of phenomena in which the microphysics may couple to the hydrodynamic flow; this might change the nature of the convection, perhaps in a "dynamic" way. Some examples are Urca processes in a convective region (see Iben 1982), "deflagration" propagation (see Nomoto, and Müller and Arnett, this volume), and neutrino-cooled burning regions.

The astrophysical "deflagration" problem is more complex than generally admitted. Consider the terminology. A "detonation" is propagated by compressive heating to ignition; a shock wave does the compression so that no microscopic mixing of matter or energy is required. A "deflagration" is propagated by conductive or radiative heating to ignition; a microscopic mixing of energy (at least) is required. In the astrophysical problem the time scales for microscopic processes to carry the heat are quite long; it is assumed that turbulent macroscopic flows carry the heat and cascade to the tiny scales on which microscopic processes can heat the unignited fuel. The nature of this process has not been understood and is only beginning to be examined (Müller and Arnett, this volume). Note that the pioneering work (Nomoto, this volume) on the nucleosynthesis yields from "deflagrating" stars may be sensitive to untested assumptions concerning the nature of the "convective" flow. In a spherical model, temperature is averaged over a spherical surface, so that one uses $T = <T>$. Nuclear reactions are sensitive to a high power of the temperature, T^n. If the temperature is not constant over the spherical surface, $<T>^n \neq <T^n>$. This is no subtle effect. In one case one might have uniformly half-burned matter (say Si), in the other half unburned (C) and half completely burned (Ni). How the total yields are affected is not known.

There is a less extreme class of phenomena in which a breakdown of geometric assumptions gives errors. As one progresses to more extreme thin-shell burning, it seems likely that different regions get out of phase, leading to non-spherically symmetric behavior, possibly affecting mixing.

E. *How does acoustic luminosity affect silicon burning?* It has been shown (Arnett 1977), both numerically and analytically, that nuclear-energized pulsations occur during silicon burning and grow to sufficient amplitude to cool by acoustic radiation (i.e., running pressure waves). Woosley (private communication) has been unable to reproduce this result numerically with the Woosley-Weaver code. In view of the sensitivity of core collapse to the initial model, this disagreement is particularly important.

For normal convection, one makes the Bousinesq approximation, which is that pressure perturbations are zero:

$$\delta P = \left.\frac{\partial P}{\partial T}\right|_v \delta T + \left.\frac{\partial P}{\partial V}\right|_t \delta V = 0 \ .$$

For δP to relax to zero, a time τ is required such that $\tau \gg \tau(\text{sound travel})$, or $v \ll s$, that is, subsonic flow. Compare this to the rise time for the rate of nuclear energy generation, ϵ. If we have

$$\frac{\partial \epsilon}{\partial T} \approx n \frac{\epsilon}{T}$$

and a heat content which is $\left.\frac{\partial E}{\partial T}\right|_V T$, then a perturbation doubles the heat content in a time

$$\tau_{\text{nuc}} \approx \frac{(\partial E/\partial T)_V}{\frac{\partial \epsilon}{\partial T}}$$

$$\approx \frac{E_T}{n\epsilon} \sim T^{-n+1} \ .$$

This rapidly decreases as temperature rises. If $\tau_{\mathrm{nuc}} \lesssim \tau$(sound travel), a hot blob expands, cooling the region by sending out pressure waves rather than by mass motion in a stately buoyant rise. As shown in Arnett (1977) this occurs during silicon burning. Further, the numerical results showed that the acoustic luminosity grew to exceed the convective luminosity, causing the convective core to decrease its size. If the trend continued, the convective core might become negligible, and silicon burning become "radiative." The core mass at collapse would approach the minimum allowable (depending on the entropy), and smaller than now found by neglecting this effect.

There seem to be two possible explanations for the discrepant results regarding acoustic radiation during silicon burning. First, Arnett used the Bodansky, Clayton, and Fowler (1968) approximate expression for energy release by quasi-equilibrium silicon burning; this might be in error for variations in temperature over a pulsational period. The error would need to be such as to cancel the pulsational driving. At first sight this might seem unlikely because quasi-equilibrium means that the time scale for emission and absorption of nucleons and alphas is fast. However, the system is sufficiently complex to require a detailed examination; this can and should be done.

The second possibility is that numerical damping in the Woosley-Weaver calculation is the culprit. The Arnett (1977) calculation was fully second order in time step δt. To incorporate a stiff system of equations for quasi-equilibrium in a convective region requires great care to maintain numerical stability. This in turn leads to an implicit numerical scheme, to backward differencing, and hence to damping.

Thus we close with another unanswered question and a further challenge to our understanding.

References

Arnett, W. D. 1969, *Ap. Spa. Sci.*, **5**, 180.

——. 1977, *Ap. J. Suppl.*, **35**, 145.

Bodansky, D., Clayton, D. D., and Fowler, W. A. 1968, *Ap. J. Suppl.*, **16**, 299.

Carr, B., Bond, R., and Arnett, W. D. 1984, *Ap. J.*, **277**, 445.

Iben, I. 1982, *Ap. J.*, **253**, 248.

Jin, L., and Arnett, W. D. 1985, *Ap. J.*, submitted.

Smarr, L., ed. 1979, *Sources of Gravitational Radiation* (Cambridge: Cambridge University Press).

Weaver, T., Woosley, S. E., and Fuller, G. M. 1983, in *Numerical Astrophysics: In Honor of J. R. Wilson,* ed. R. Bowers (Portola, Calif.: Science Books International).

Woosley, S. E., and Weaver, T. 1982, in *Supernovae: A Survey of Current Research,* ed. M. J. Rees and R. J. Stoneham (Dordrecht: Reidel).

10

Models of Type II Supernova Explosions

W. Hillebrandt and R. G. Wolff

10.1 Introduction

Among the various explanations given of why some stars explode as supernovae, the following scenarios are presently favored by most researchers in the field. The progenitors of Type I events are believed to be white dwarfs which are disrupted by thermonuclear explosions, the energy source being explosive carbon burning (see, e.g., Müller and Arnett, Woosley, and Nomoto, this volume). In most models it is assumed that the white dwarf grows in mass beyond the Chandrasekhar limit by accreting matter from a nearby companion star and then contracts and ignites its nuclear fuel. Although it has not yet been determined whether binary evolution does indeed in some cases lead to such a situation, these models are able to explain both the light curves (Arnett 1982; Woosley, Weaver, and Taam 1980) and the spectra (see, e.g., Branch 1984 and this volume) and therefore are generally believed to be the correct explanation.

Type II supernovae, on the other hand, are thought to be the outcome of the collapse of the cores of massive stars, leaving behind a compact remnant, namely a neutron star. In this picture the energy observed in the explosion must ultimately come from the gain in gravitational binding of the core. Two mechanisms that are potentially able to transform a small fraction of the gravitational energy into outward momentum of the stellar envelope are discussed, neither of which, however, so far gives a satisfactory explanation of the entire phenomenon.

In one class of models the stellar envelope is ejected by a hydrodynamical shock wave generated by the rebounding inner core near nuclear matter density, but this mechanism seems to work only for a very limited range of precollapse stellar models. In fact, one can show quite generally that the crucial quantity which determines whether the shock wave can propagate successfully out to the stellar envelope or will turn into a standing accretion shock is the difference between the mass of the unshocked inner part of the core and the mass of the original "iron" core. If this difference exceeds approximately 0.5 $M_\odot$, damping of the shock by nuclear photodissociations becomes so severe that no explosion results. Both the mass and the entropy of the iron core at the onset of collapse are sensitive to details of nuclear reaction rates, in particular to electron capture rates during silicon burning as well as to the specific way convection during hydrostatic stellar evolution is modeled (Arnett, this volume; Weaver et al. 1982) for stars more massive than about 10 $M_\odot$ on the main sequence. Typical values given presently in the literature range from $M_{core} \approx 1.35\ M_\odot$ to 1.6 $M_\odot$, where the smaller masses seem to be more realistic (Weaver et al. 1982). For stars with masses around 9 $M_\odot$ the core mass may be even smaller (Nomoto 1983). Central entropies are found to be around 1 k_B/nucleon, somewhat smaller numbers not being excluded (Arnett, this volume; Weaver et al. 1982). Low entropies again favor explosions, since the lepton fraction in the collapsing core increases with decreasing entropy, and therefore the mass of the unshocked core also increases.

The main aim of this paper is to demonstrate the sensitivity of computer simulations of stellar collapse and the subsequent shock propagation phase to details of the initial models as well as of the numerical code. It will be shown that in particular the initial entropy of the core is of crucial importance. From our numerical studies we conclude that even for small iron core masses ($M_{core} \approx 1.35\ M_\odot$) initial entropies of at most 0.7 k_B/nucleon are required for successful explosions. But even for such low entropies, we have so far found only a marginal result, which calls for further improvements of both the numerical treatment and the microphysics input data. If our results should then remain qualitatively correct, alternative scenarios, such as "late time" explosions mediated by neutrinos diffusing out of the proto-neutron star (Wilson 1983), should be investigated in more detail.

10.2 The Equation of State

Because the deviation of the effective adiabatic index at low entropies from the critical value 4/3 together with the fraction of trapped leptons determines the mass inside the homologously collapsing part of the core and thus the mass zones where the shock forms, one of the most important pieces of input physics for any computer simulation of stellar collapse and supernova explosions is the equation of state. In addition, the stiffness of the equation of state at high entropies strongly influences the propagation of the shock wave. Finally, the density at which the collapse is stopped and reversed, and consequently the energy that is originally put into the shock, depends upon details of the equation of state at the transition from a mixture of nuclei embedded in a fluid of nucleons to homogenous nuclear matter. In particular, the possible existence of a bubble phase may significantly lower the adiabatic index there.

To obtain a good qualitative description of the equation of state is certainly a difficult task and requires the adoption of appropriate many-body techniques to treat the nuclear interactions in a realistic manner. Among the available methods the variational methods of Friedman and Pandharipande (1981) so far give the most satisfactory results concerning homogenous nuclear and neutron matter. Because of technical difficulties, this method has not yet been applied to inhomogenous nuclear matter.

At present the only tractable method seems to be the so-called temperature dependent Hartree-Fock method (Bonche and Vautherin 1981) and simplifications of it like the Thomas-Fermi model (Marcos et al. 1982; Barranco and Buchler 1980; Suraud and Vautherin 1983; Ogasawara and Sato 1982) or the compressible liquid drop model (Baym et al. 1971) and extensions of it to finite temperatures (Lamb et al. 1978, 1981, 1983; Lattimer 1981; Ravenhall et al. 1983; Pethick et al. 1983).

By using vector techniques on a Cray-1 computer, self-consistent Hartree-Fock computations of the equation of state may be as fast as solving Thomas-Fermi equations, where additional ambiguities concerning the kinetic energy functional come into play. We have therefore decided to perform Hartree-Fock computations and to tabulate the equation of state in a sufficiently dense grid in ρ, T (or S), and Y_e for use in hydrodynamical codes. Additional advantages of the Hartree-Fock approach are that at low entropies the influence of shell closures is a natural outcome of the model and that from both bound and continuum single-particle states based on Hartree-Fock potentials, nuclear level densities can be constructed for unknown nuclei by using the Grand Partition Function approach (Huizenga and Moretto 1972;

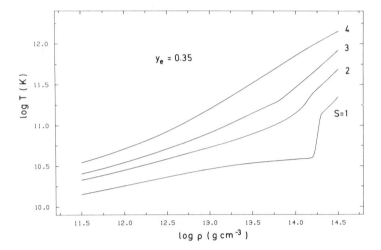

Fig. 10.1. Isentropes for a fixed electron fraction $y = 0.35$ resulting from mean field computations using Skyrme force parametrizations due to Köhler.

Wolff 1980). The latter are needed for the computation of multicomponent mixtures of nuclei in nuclear statistical equilibrium at lower densities $(\rho \lesssim 10^{12} \text{ g cm}^{-3})$.

Before we describe in more detail the method used by us, we want to mention a serious shortcoming of all computations that have so far been performed. In the conventional methods the cellular structure of matter is approximated by spherical Wigner-Seitz cells. This approximation, however, is only essentially exact as long as the radius of the inhomogeneities (nuclei) is small compared to the cell radius and the concentration of matter in regions close to the cell boundary is small. Obviously, these approximations become questionable near nuclear matter density, where the nuclear radii become comparable to the radius of the Wigner-Seitz cells. Also, the existence of a bubble phase found in some calculations just below nuclear matter density may be an artifact of this method. A better treatment of the transition region requires that nonspherical effects such as a real lattice structure be taken into account, because close to phase transitions, as noted by Landau (1932), small effects become essential (Lamb et al. 1983). Calculations which explicitly treat three-dimensional modifications of the conventional method are currently in progress and will be discussed elsewhere.

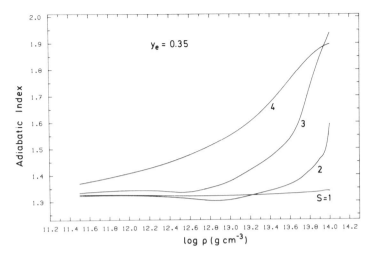

Fig. 10.2. Adiabatic indices at subnuclear densities as predicted by mean field calculations as in figure 10.1.

In order to obtain the equation of state displayed in figures 10.1 to 10.4, which has been used in the numerical studies described in the following sections, we started from a model Hamiltonian, which for the nucleons in second quantization is written

$$H = \sum_{i,j} <i|t|j> a_i^+ a_j + \frac{1}{2} \sum_{i,j,k,l} <ij|V|kl> a_i^+ a_j^+ a_k a_l \quad , \tag{1}$$

where t is the kinetic energy operator, V is an effective nucleon-nucleon interaction, and a_i^+ and a_i are nucleon creation and annihilation operators. The thermal Hartree-Fock method is based on the following form of the one-body density operator:

$$\hat{\rho} = \frac{1}{Z} \exp\left\{ -\sum_i \alpha_i a_i^+ a_i \right\} \quad , \tag{2}$$

$$Z = \sum_{n_1, n_2, \dots = 0, 1} <n_1 n_2 \dots | \exp\left\{ -\sum_i \alpha_i a_i^+ a_i \right\} | n_1 n_2 \dots> \quad ,$$

where the sum $\sum_{n_1, n_2 \dots}$ goes over all completely antisymmetrized states and the α_i are parameters which will be determined from the variational principle. The expression for Z, the Grand Partition Function, reduces to

$$Z = \prod_i \left(1 + e^{-\alpha_i}\right) \quad . \tag{3}$$

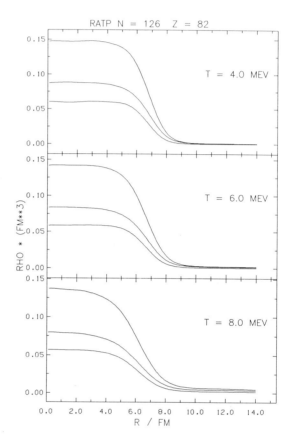

Fig. 10.3. Proton, neutron, and total baryon density distributions within a Wigner-Seitz sphere of radius $R = 14$ fm, at temperatures of relevance in stellar collapse. The density profiles are based on mean field calculations using the Skryme force parameters given by Rayet et al. (RATP). Note the remarkable difference in the saturation value of the density.

The energy is now given by

$$E = Tr\,(\hat{\rho} \cdot H) = \sum_i <i|t|i> n_i + \frac{1}{2} \sum_{i,j} n_i n_j <ij|V|ij - ji> \; , \qquad (4)$$

where the n_i are defined by

$$n_i = Tr\,(\hat{\rho}\, a_i^+ a_i) = (1 + e^{\alpha_i})^{-1} \qquad (5)$$

and have the usual meaning of occupation probabilities for single-particle

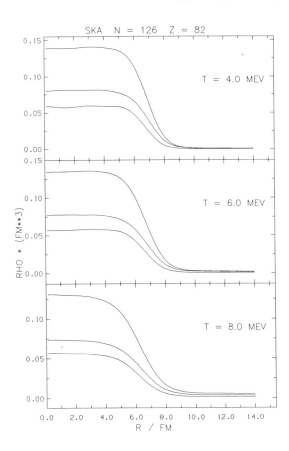

Fig. 10.4. Same as figure 10.3 for force parameters Ska.

states i. The entropy is given by

$$S = -Tr\left(\hat{\rho}\ln\hat{\rho}\right) = -Tr\left[\hat{\rho}\cdot\left\{-\sum_i \alpha_i a_i^+ a_i - \ln Z\right\}\right]$$

$$= -\sum_i \left\{n_i \ln n_i + (1-n_i)\ln(1-n_i)\right\} \ , \tag{6}$$

and, finally, the particle number is evaluated from

$$N_k = Tr\left(\hat{N}_k\hat{\rho}\right) = \sum_i n_i^k \ , \tag{7}$$

where the index k denotes either neutrons or protons.

The Grand canonical potential can now be computed from

$$G = G(T,\Omega,\mu_k) = F - \mu_k N_k = E - T \cdot S - \mu_k N_k \ .$$ (8)

Here Ω is the volume and μ_k are neutron and proton chemical potentials, respectively. The minimization of G with respect to the parameters α_i and the single-particle wave functions at fixed baryon number then yields

$$\frac{\delta G}{\delta \phi_m} = 0 \Rightarrow W|\phi_m> = e_m|\phi_m> \ ,$$ (9)

where $<i|W|j> = <i|t|j> + \sum_l n_l\{<i|V|jl - lj>\}$ are the matrix elements of the Hartree-Fock Hamiltonian and e_m are the single-particle energies and

$$\frac{\delta G}{\delta \alpha_m} = 0 \Rightarrow n_m = \{1 + \exp[(e_m - \mu_k)/k_B T]\}^{-1} \ .$$ (10)

For given temperature T, neutron and proton number N_k, and volume of a Wigner-Seitz cell Ω_c, the Grand canonical potential can easily be evaluated once the self-consistent solution of the Hartree-Fock equations (9) is known. Equation (10) then determines the neutron and proton chemical potentials. Finally, the pressure is obtained from

$$p = -\frac{\partial G}{\partial \Omega} \ .$$ (11)

It is, of course, straightforward to include in the above description other fermions such as electrons or positrons. In our computation proton-proton, proton-lepton, and lepton-lepton Coulomb interactions were included. In addition, contributions from electron-positron pairs were also taken into account. Neutrinos, on the other hand, were treated separately because their number is in general not locally conserved.

In the hydrodynamical studies discussed later, the effective nucleon-nucleon interaction was a Skyrme-type force with parameters given by Köhler (1976), which were determined by fitting properties of nuclear matter such as saturation density, saturation energy, compressibility, and symmetry energy, as well as properties of finite nuclei. This choice is, of course, not unique, and other parametrizations may be superior (see, e.g., Rayet et al. 1982). The Köhler force leads to a rather stiff equation of state (see also table 10.1 for a comparison of different forces), which in turn favors successful shock propagation. Unsuccessful models therefore will not change qualitatively if another force leading to a softer equation of state is used.

Table 10.1

Thermodynamical Properties of Pb at Different Temperatures but Fixed Density Resulting from Thermal Hartree-Fock Equations

T/MeV	RATP			Ska			SkM		
	U/A	S/A	P	U/A	S/A	P	U/A	S/A	P
0.0	−7.82	0	−0.086	−7.59	0	−0.010	−7.89	0	−0.091
2.0	−7.43	0.31	−0.063	−7.20	0.29	−0.064	−7.44	0.34	−0.054
4.0	−6.24	0.71	−0.056	−6.03	0.69	−0.057	−6.15	0.77	−0.049
6.0	−3.63	1.25	−0.041	−3.48	1.22	−0.041	−3.42	1.34	−0.034
8.0	1.08	1.96	−0.011	1.15	1.92	−0.010	1.43	2.07	−0.008

Note: The internal energy U is given in MeV per baryon, the entropy is given in units of k-Boltzmann and baryon, and the pressure is given in MeV/fm^3. Computations have been performed for Wigner-Seitz spheres of radius 14 fm and Skyrmeforce parametrizations given by Rayet et al. (RATP), Köhler (Ska), and Bohigas et al. (SkM). At low temperatures internal energies based on RATP agree with energies resulting from SkM; however, at high temperatures RATP and Ska evidently show smaller internal energies and entropies than SkM.

In order to solve the Hartree-Fock equations, boundary conditions for the single-particle wave functions have to be specified. Assuming spherical symmetry for the Wigner-Seitz cell, single-particle states can be labeled by their angular momentum l. Boundary conditions consistent with periodicity are then

$$\left. \frac{d\phi_i}{dr} \right|_{r=R_c} = 0 \qquad \text{for odd } l_i \ ,$$

$$\phi_i(r = R_c) = 0 \qquad \text{for even } l_i \ . \tag{12}$$

These boundary conditions do not guarantee that the density is also periodic. In fact, in particular for bubble-like phases, which give the lowest free enthalpy at densities between about 0.5 ρ_0 and ρ_0 (ρ_0 = nuclear matter density), the density may show nonzero gradients at the cell boundary. This finding indicates a breakdown of the assumption of spherical symmetry and calls for self-consistent band structure calculations. Such calculations are in progress and will be presented elsewhere. Preliminary results show, however, that a bubble-like structure is rather unlikely. For the equation of state

Table 10.2
Properties of Pre-Supernova Stellar Models of Different
Main Sequence Masses

Properties	Main Sequence Mass		
	~9 $M_\odot$	10 $M_\odot$	20 $M_\odot$
He core mass	2.2–2.4	2.7	~5
Nondegenerate burning up to a core of:	O+Ne+Mg (1.28 $M_\odot$)	Ne+Si (1.51 $M_\odot$)	Fe+Ni (1.36 $M_\odot$)
Degenerate burning	O → "Fe" (0.7 $M_\odot$)	Ne → "Fe" (1.36 $M_\odot$)	...
Collapse triggered by:	e^- captures on Mg, Ne, Na, ..., "Fe"	e^- captures on Ne, S, ...	e^- captures, photodisintegration
Central entropy (initially)	0.9	0.85	0.7 (1.1)
Electron concentration Y_e (initially)	0.41	0.43	0.42

displayed in figures 10.1 to 10.4, only solutions with maximum density at the center of the cell have been chosen, since they seem to be a better approximation of the true solution of the problem.

10.3 Initial Models and the Core Collapse Phase

The essential properties of several precollapse stellar models that have been investigated by us over the last few years are summarized in table 10.2. The parameter that differentiates between the various models is their main sequence mass or He core mass after core H burning.

Models with He core masses between 2.2 and 2.4 $M_\odot$ evolve into degenerate O-Ne-Mg cores of about 1.3 $M_\odot$ (Nomoto 1983; Miyaji et al. 1980). These cores contract rapidly on the time scale of electron captures on ^{24}Mg and ^{20}Ne, and oxygen burning is ignited once the central density exceeds 2.5

$\times$ 10^{10} g cm^{-3}. A deflagration front forms which incinerates the matter into a nuclear statistical equilibrium (NSE) composition. When the NSE core contains 0.7 $M_{\odot}$, further evolution proceeds on a hydrodynamic time scale. By the time the central density exceeds nuclear matter density, the mass inside the O-burning front is still rather small, namely 1.1 $M_{\odot}$.

Stars with He core masses of about 2.7 $M_{\odot}$, on the other hand, burn their nuclear fuel nondegenerately until a Ne-Si core of 1.51 $M_{\odot}$ is formed (Woosley et al. 1980). Degenerate Ne burning then transforms 1.36 $M_{\odot}$ of the core into iron-group nuclei, and the collapse is again triggered by e^{-} captures. In this case the central entropy is slightly lower and the electron concentration at neutrino trapping is slightly higher than in the previous models.

More massive stars, finally, burn all their nuclear fuel under nondegenerate conditions. As an example of this class of models we have chosen a 20 $M_{\odot}$ star like that whose evolution from the main sequence has been modeled by Weaver et al. (1982). It differs from those modeled previously in the literature by its rather small iron core mass and its very low central entropy at the onset of collapse. This model therefore seems to be a good candidate for a core-bounce supernova explosion.

Before we discuss the results of our numerical studies, we should mention a consistency problem which essentially all computer simulations of stellar collapse encounter. Stellar evolution codes predict the properties of stars, such as density, pressure, temperature (or entropy), and composition, at the onset of collapse, which can then in principle be used as input data for hydrodynamical simulations. In practice, however, the equations of state used in the various approaches are not identical, and one therefore is left with some ambiguities in choosing an initial model.

One can, for example, take density, pressure, and composition (electron concentration in NSE) from the evolution code and compute the temperature (entropy) from the equation of state that is used in the hydrodynamics. This particular choice guarantees that the acceleration terms will fit but may lead to significant differences in the temperature (entropy). An alternative choice is to take the entropy from the evolution model and compute the pressure from the equation of state. Although these differences might look unimportant at first glance, in actuality they are not. Weaver et al. (1982), for example, do not consider Coulomb lattice correlations in their equation of state at high densities. For a density of 10^{10} g cm^{-3}, a temperature of 7.7 $\times$ 10^{9} K,

Table 10.3
Properties of Supernova Models at Core Bounce

Property	$\sim 9\ M_\odot$	$10\ M_\odot$	$20\ M_\odot$
$M_{\mathrm{HC}}/M_\odot$	0.68	0.81	0.85 (0.7)
Y_e^c	0.29	0.32	0.32 (0.30)
Y_ν^c	0.06	0.08	0.09 (0.07)
$S^c\ (k_\mathrm{B}/\text{nucleon})$	1.1	1.2	1.1 (1.5)
$\rho^c\ (\text{g cm}^{-3})$	2.9×10^{14}	3.5×10^{14}	3.3×10^{14} (3.0×10^{14})
$T^c\ (\text{K})$	2.8×10^{11}	3.2×10^{11}	2.4×10^{11} (3.1×10^{11})
$E_{\text{shock}}^i\ (10^{51}\ \text{ergs})$	~ 7	~ 8	~ 8 (~ 7)

Note: M_{HC} is the mass of the homologous core. Central values of different quantities are given. E_{shock}^i is the energy given initially to the shock. Numbers in parentheses are obtained for the 20 $M_\odot$ model with high initial entropy ($S = 1.1\ k_\mathrm{B}/\text{nucleon}$).

and $Y_e = 0.42$, these corrections result, however, in a pressure reduction of about 3%. If we therefore select density, pressure, and Y_e from the stellar model and compute from our equation of state the entropy in the innermost mass zone of their star, it is raised from their value of 0.7 to 1.1 $k_\mathrm{B}/\text{nucleon}$. And, as we shall demonstrate later, the model with the low entropy is close to an explosion, whereas the one with higher entropy is not.

A comparison of the properties of the various models at core bounce is given in table 10.3. The properties are remarkably similar, despite the fact that different initial models, and in the 10 $M_\odot$ case a different equation of state as well, have been used (see, e.g., Hillebrandt 1982). In particular, the energies given initially to the shock are almost the same for all computations. The only obvious differences show up in the mass of the homologous core, which is determined essentially by the lepton concentration at neutrino trapping. This latter quantity depends on both the initial entropy of the core and the collapse time scale. The 9 $M_\odot$ model has a rather low lepton fraction ($Y_L \approx 0.35$) because during the early "slow" contraction phase most of the neutrinos produced by e^- captures still escape from the core and the contraction is governed by the e^- capture time scale.

For the more massive stars, on the other hand, the final lepton concentration is determined mainly by the initial entropy, and because of the higher proton concentration models with higher entropies end up with smaller lepton concentrations and consequently smaller homologous cores.

In the case of the same stellar model (20 $M_\odot$) but different initial entropies, this difference between the two homologous cores is about 0.15 $M_\odot$. Consequently, for the model with higher entropy an additional energy of about 2×10^{51} ergs has to be paid by the shock for nuclear photodissociations relative to the low entropy case. Since the shock energy is also somewhat lower in the first case, quantitative and even qualitative differences between the two simulations can be expected.

10.4 Supernova Explosions: Yes or No?

Since a discussion of the 9 and 10 $M_\odot$ models has been given elsewhere (Hillebrandt 1982, 1984; Hillebrandt et al. 1984), we do not present the details of the models here but rather summarize briefly why explosions were found in these cases.

The 9 $M_\odot$ model gave rise to a very energetic explosion ($E_{ex} \approx 2 \times 10^{51}$ ergs) because the iron core mass was very small and, owing to the presence of an oxygen-burning shell, the infall velocities of the outer parts of the core (and therefore the ram pressure imposed on the shock) were small, too. Both effects were responsible for the fact that the shock did not experience much damping on its way out (see figure 10.5). Numerical damping, which is usually caused by the artificial smearing of the shock over several mass zones due to pseudoviscosity, was also kept small by an appropriately good zoning. Because thermal emission of neutrinos was neglected, the explosion energy is certainly overestimated, but it is unlikely that the general conclusions would

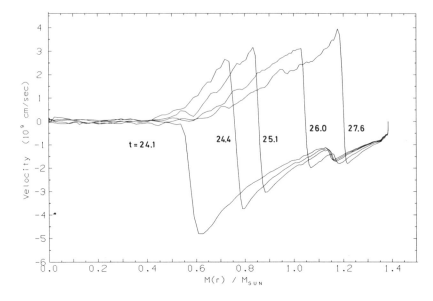

Fig. 10.5. Velocity profiles of a 9 $M_\odot$ star at different times. Inward velocities are negative; outward velocities are positive. The curves are labeled with the time in ms measured from the beginning of the computations. The shock reaches the O-burning shell (at about 1.1 $M_\odot$) 2.3 ms after core bounce.

have to be changed if this additional energy loss were also included.

The 10 $M_\odot$ model, on the other hand, gave only a very weak explosion ($E_{ex} \approx 5 \times 10^{50}$ ergs) (Hillebrandt 1982). Although in this case the mass of the homologous core was larger and the initial shock energy was greater than in the previous model, these properties were more than compensated for by the larger iron core mass and the higher ram pressure. It was only thanks to the steep density gradient at the edge of the iron core that a marginal explosion resulted at all (see figure 10.6).

Again, we should mention a few shortcomings of this particular computation. First, the equation of state based on a simple Saha equation approach that was used does not correctly describe the nuclear properties near nuclear matter density. The more realistic Hartree-Fock method discussed in § 10.2 predicts a significantly stiffer equation of state above 10^{13} g cm^{-3} and is therefore more favorable for explosions. Neutrino transport, on the other hand, has been treated in a very crude fashion, and neutrino losses from the shock were certainly underestimated. Whether a more realistic simulation of the

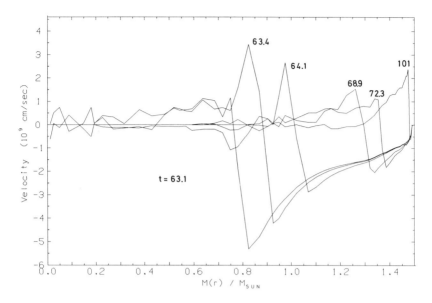

Fig. 10.6. Same as figure 10.5 but for the 10 $M_\odot$ model. In these computations the shock accelerates when it reaches the edge of the iron core about 16 ms after core bounce (at about 1.4 $M_\odot$).

collapse of this particular model will confirm the earlier results is thus an open question, which will be investigated in the near future.

From the arguments given above and our experience with previous core collapse computations, we would expect that the 20 $M_\odot$ model with "high" initial entropy would not explode, because the core structure was similar to the 10 $M_\odot$ model but both the mass of the homologous core and the initial shock energy were smaller. And, in fact, we found that the shock changed into a standing accretion shock at 1.15 $M_\odot$ and a radius of 110 km (see table 10.4 and figure 10.7). By improving the treatment of pseudoviscosity (which caused too much numerical damping in the first simulation) and by redistributing the mass zones somewhat, we obtained a shock propagating to 150 km ($\approx$ 1.20 $M_\odot$) but then again stagnating. These results are in good agreement with what one expects from analytical considerations (Mazurek 1982; Burrows and Lattimer 1982).

The only promising candidate for an explosion of a more massive star was therefore the 20 $M_\odot$ model with low initial entropy. But for this model also we have not as yet been able to obtain a definite result. In order to keep

Table 10.4
Summary of the Results of Various Computations

Model	E_{ex} (ergs)	$M_{N^*}/M_\odot$	Remarks
$\sim 9\ M_\odot$	$\sim 2 \times 10^{51}$	~ 1.2	Thermal neutrinos neglected
$10\ M_\odot$	$\sim 5 \times 10^{50}$	~ 1.4	Simple equation of state, poor treatment of neutrino transport
$20\ M_\odot$ $S_i^c = 1.1$	. . .	. . .	Shock to 110 km (1.15 $M_\odot$)
$20\ M_\odot$ $S_i^c = 1.1$	. . .	. . .	Improved zoning, pseudoviscosity, shock to 150 km (1.20 $M_\odot$)
$20\ M_\odot$ $S_i^c = 0.7$	$< 2 \times 10^{50}$	> 1.4	Zoning problems ($m(r) > 1.2\ M_\odot$) shock to > 350 km ($> 1.35\ M_\odot$)

Note: E_{ex} is the explosion energy in the model computation with 9 $M_\odot$ and 10 $M_\odot$, respectively. For the low entropy 20 $M_\odot$ model the energy in the shock at the end of the computation is given. M_{N^*} is the mass of the neutron star that is left over.

numerical noise as low as possible, this model computation was performed with 100 zones of equal mass $\Delta m \approx 0.017\ M_\odot$. As can be seen in figure 10.8, this approach gave a good resolution of the shock out to about 1.15 $M_\odot$. Since our treatment of pseudoviscosity smeared the shock artificially over about 10% of the radius and since the spacing of the zone just ahead of the shock then became significantly larger, we observed numerical damping due to bad zoning. Nevertheless the shock did propagate to a radius of 350 km ($\approx$ 1.35 $M_\odot$) before we stopped the computations, since the shock speed had dropped below 10^9 cm s^{-1} and at best a marginal explosion would be expected.

A comparison with the 10 $M_\odot$ model mentioned earlier shows that the shock propagation is quite similar in both computations out to a mass of

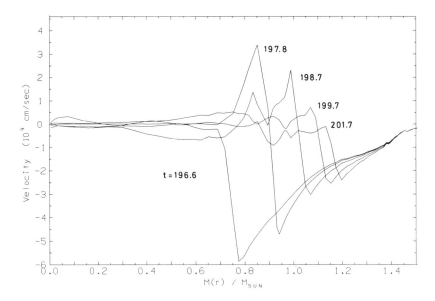

Fig. 10.7. Same as figure 10.5 but for the 20 $M_\odot$ model with high initial entropy. In this model computation, strong damping due to nuclear photodissociations and neutrino losses causes shock stagnation at a radius of 110 km (about 1.15 $M_\odot$).

about 1.3 $M_\odot$, as would be expected from their similar properties at core bounce. The advantage of the slightly larger mass of the unshocked part of the inner core in the 20 $M_\odot$ model is compensated for by more severe neutrino losses caused by the more accurate treatment of neutrino transport. Only after the shock has passed the zones at 1.3 $M_\odot$ do differences become obvious. In the 10 $M_\odot$ model the shock accelerates due to the steep density gradient at the edge of the original iron core, whereas it was still weakened by nuclear photodissociations in the 20 $M_\odot$ case, which had different density and temperature structure there.

The closeness of the 20 $M_\odot$ star to an explosion makes it worthwhile to reinvestigate this particular model with improved numerical techniques, the most straightforward being an increase in the number of mass zones.

10.5 Summary and Conclusions

Our most recent computations still do not allow us to answer the question of whether there exists an upper mass limit for the progenitor stars of Type II supernova explosions of around 10 to 15 $M_\odot$, since even the most

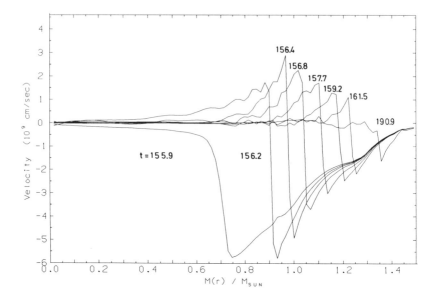

Fig. 10.8. Same as figure 10.5 but for the 20 $M_\odot$ model with low initial entropy. This time the shock propagates significantly farther out as compared to the same model with high initial entropy. But at best a marginal explosion can be expected.

favorable stellar evolution models of more massive stars available to date do not show clear evidence for an explosion caused by the core-bounce mechanism.

We can confirm, however, our earlier conclusions that a small iron core mass and low initial entropy favor explosions. These numerical results are also in good agreement with analytical considerations.

Although from the observed Type II supernova rate there is no need for explosions of more massive stars and also direct observational evidence is rather weak, nucleosynthesis arguments still favor the idea that at least some massive stars explode by the core-bounce mechanism, thereby producing most of the elements from silicon to iron (Woosley and Weaver 1982). If we adopt this point of view and assume for the moment that improvements in the numerical treatment will not change the results given here, we are left with the problems that either the initial stellar models used in hydrodynamical simulations are still not correct or the simple one-dimensional models do not yield the right answer. Concerning the first problem it is hard to imagine that stellar evolution models will predict still significantly smaller core masses and entropies. The low entropy 20 $M_\odot$ model used in our present study seems to be at the lower limit that can be expected for these quantities. Late-time explosions generated by neutrinos leaking out of the inner core as found by Wilson (1983) will probably not resolve the nucleosynthesis problem, because Wilson's computations predict a rather weak shock and a mass cut outside the original iron core.

It therefore seems preferable to abandon the assumption of spherical symmetry and to reconsider rotating stellar models. Here a rather stiff equation of state, such as the one presented in § 10.2, may also qualitatively change the results obtained previously by Müller and Hillebrandt (1981). Since the adiabatic index from this equation of state is only slightly below 4/3 for most densities of interest, already a small amount of angular momentum will raise the effective γ to values above 4/3, thus changing the dynamics of core collapse considerably. One can even imagine that a large amplitude bounce at densities much below nuclear matter density may result.

References

Arnett, W. D. 1982, *Ap. J.*, **253,** 785.

Barranco, M., and Buchler, J. R. 1980, *Phys. Rev.*, **C22,** 1729.

Baym, G., Bethe, H. A., and Pethick, C. J. 1971, *Nucl. Phys.*, **A175,** 225.

Bohigas, O., Krivine, H., and Treiner, J. 1980, *Nucl. Phys.*, **A336,** 155.

Bonche, P., and Vautherin, D. 1981, *Nucl. Phys.*, **A372,** 496.

Branch, D. 1984, in *Proc. XI Texas Symposium on Relativistic Astrophysics (Ann. N.Y. Acad. Sci.,* **422,** in press).

Burrows, A., and Lattimer, J. M. 1983, *Ap. J.*, **270,** 735.

Friedman, B., and Pandharipande, V. R. 1981, *Nucl. Phys.*, **A391,** 502.

Hillebrandt, W. 1982, *Astr. Ap.*, **110,** L3.

———. 1984, in *Proc. XI Texas Symposium on Relativistic Astrophysics (Ann. N.Y. Acad. Sci.,* **422,** 197).

Hillebrandt, W., Nomoto, K., and Wolff, R. G. 1984, *Astr. Ap.*, **133,** 175.

Huizenga, J. R., and Moretto, L. G. 1972, *Ann. Rev. Nucl. Sci.*, **22,** 427.

Köhler, H. S. 1976, *Nucl. Phys.*, **A258,** 301.

Lamb, D. Q.; Lattimer, J. M.; Pethick, C. J.; and Ravenhall, D. G. 1978, *Phys. Rev. Letters,* **41,** 1623.

——. 1981, *Nucl. Phys.*, **A360,** 459.

——. 1983, *Nucl. Phys.*, **A411,** 449.

Landau, L. D. 1937*a*, *JETP*, **7,** 19.

——. 1937*b*, *JETP*, **7,** 627.

Lattimer, J. M. 1981, *Ann. Rev. Nucl. Part. Sci.*, **31,** 337.

Lattimer, J. M., and Ravenhall, D. G. 1978, *Ap. J.*, **223,** 314.

Marcos, S., Barranco, M., and Buchler, J. R. 1982, *Nucl. Phys.*, **A381,** 507.

Mazurek, T. J. 1982, *Ap. J. (Letters)*, **258,** L13.

Miyaji, S.; Nomoto, K.; Yokoi, K.; and Sugimoto, D. 1980, *Pub. Astr. Soc. Japan*, **32,** 303.

Müller, E., and Hillebrandt, W. 1981, *Astr. Ap.*, **103,** 358.

Nomoto, K. 1983, *Ap. J.*, in press.

Ogasawara, R., and Sato, K. 1982, *Prog. Theor. Phys.*, **68,** 222.

Pethick, C. J., Ravenhall, D. G., and Lattimer, J. M. 1983, *Phys. Letters,* **B128,** 137.

Ravenhall, D. G., Pethick, C. J., and Lattimer, J. M. 1983, *Nucl. Phys.,* **A407,** 571.

Ravenhall, D. G., Pethick, C. J., and Wilson, J. R. 1983, *Phys. Rev. Letters,* **50,** 2066.

Rayet, M.; Arnould, M.; Tondeur, F.; and Paulus, G. 1982, *Astr. Ap.*, **116,** 183.

Suraud, E., and Vautherin, D. 1984, preprint IPNO/TH 83-60.

Weaver, T. A., Woosley, S. E., and Fuller, G. M. 1982, *Bull. AAS,* **14,** 957.

Wilson, J. R. 1983, preprint.

Wolff, R. G. 1980, diplomathesis, TH Darmstadt.

Woosley, S. E., and Weaver, T. A. 1982, in *Supernovae: A Survey of Current Research,* ed. M. J. Rees and R. J. Stoneham, p. 79 (Dordrecht: Reidel).

Woosley, S. E., Weaver, T. A., and Taam, R. E. 1980, in *Type I Supernovae,* ed. J. C. Wheeler, p. 164 (Austin: University of Texas Press).

11

Hydrostatic Nucleosynthesis
and Pre-Supernova Abundance Yields
of Massive Stars

Friedrich-K. Thielemann and W. David Arnett

Abstract

The thermonuclear synthesis of nuclei during quasi-hydrostatic stages of the evolution of massive stars is examined in detail. Our network is essentially complete up to $A = 74$. We formulate the equations for the evolution of temperature and density in a simple way (assuming polytropic structure of the stellar core), so that we examine the nucleosynthesis in a manner we hope to be insensitive to particular assumptions regarding the nature of stellar convection and mixing.

We present the model and results of core He- and C-burning for original He-core masses of 1.5, 2, 4, 8, and 16 $M_\odot$.

Detailed core He-burning results are needed to estimate nucleosynthesis yields of massive stars. The influence of a faster $^{12}C(\alpha,\gamma)$ ^{16}O rate on He-burning products is studied systematically. The onset and conditions of s-processing in core He- and C-burning are also examined.

The study of later core burning stages gives information about minimum size reaction networks which are able to predict accurately the neutron excess $\eta(t) = \sum_i (N_i - Z_i) Y_i$, one of the most important quantities in final burning stages of massive stars before the onset of core collapse. Neutronization due to electron capture on nuclei in O- and Si-burning and its consequences for the link between the two quasi-equilibrium (QSE) groups around Si and Fe in Si-burning is analyzed in detail.

Shell burning conditions are simulated, and *pre*-supernova nucleosynthesis yields of a 20–25 $M_\odot$ star (8 $M_\odot$ original He-core) are predicted. Varying the $^{12}C(\alpha,\gamma)$ ^{16}O rate within the present uncertainties seems to remove problems with type II supernova abundance yields (compared to solar) which have existed in recent years.

These conclusions need to be checked in a consistent stellar evolution and explosive nucleosynthesis calculation.

11.1 Introduction

While explosive nucleosynthesis calculations have been performed with large reaction networks and analyzed in detail (see, e.g., Arnett and Truran 1969; Howard et al. 1972; Woosley, Arnett, and Clayton 1973; Morgan 1980; Thielemann, Arnould, and Hillebrandt 1979; Arnould et al. 1980; Woosley and Weaver 1982), hydrostatic nucleosynthesis during stellar evolution has been treated in a less detailed manner. The most advanced calculation in this respect (Weaver, Zimmerman, and Woosley 1978; Woosley and Weaver 1982) used a reaction network containing 19 nuclear species up to oxygen burning and switched over to an equilibrium network in silicon burning. More recently Weaver, Woosley, and Fuller (1983) modified and enlarged this equilibrium network.

At the present time and with the present computers, a full stellar evolution calculation including a complex nuclear reaction network is not yet feasible. We have used a simple (one-zone) approximation of polytropic structure for the convective core, to simulate central conditions (temperature and density) during subsequent evolutionary stages of core burning. All energy gain and loss terms were included (nuclear energy generation, neutrino losses, radiation losses, and neutrino losses from β-decay and electron capture on nuclei).

The intent is to explore with such a simple (stellar) model the nucleosynthesis in hydrostatic burning phases in detail and to determine the minimum size reaction networks for use in stellar evolution codes, which still keep accurate track of the neutron excess $\eta(t) = \sum_i (N_i - Z_i) Y_i$ (or the electron abundance $Y_e = (1 - \eta)/2$). This quantity is of major importance in late burning stages and the final collapse of cores of massive stars before a type II supernova explosion. As η is of the order of 10^{-2}, nucleosynthesis flows have to be considered which are less than 0.01 of major flows (which govern energy generation).

After a description of the network (§ 11.2) and the stellar model (§ 11.3), we discuss core He- and C-burning nucleosynthesis in detail (§§ 11.4 and 11.5). The products of core He-burning will be present in the debris ejected in type

II supernova events as the outer part of the He-burning core, and will not be affected by later burning stages. Nucleosynthesis itself is of general interest in this respect and also with respect to s-processing in core He-burning as compared to earlier studies (Couch et al. 1973; Lamb et al. 1977). The influence of recent indications of $^{12}C(\alpha,\gamma)\,^{16}O$ rate enhanced by a factor of 2 to 5 (Kettner et al. 1982; Langanke and Koonin 1983) is analyzed and found to be of major importance for the $^{12}C/^{16}O$ ratio at the end of He-burning and the fuels of subsequent C- and O-burning.

In C-burning the convective core is of the size of the Chandrasekhar mass (as is the case in later stages), which is comparable to the size of the neutron star remaining after a type II supernova event. In this respect it is unclear whether core C-burning products are of interest for nucleosynthesis if no mixing mechanism removes material from this core. For less massive type II supernova progenitors ($M \approx 15\ M_\odot$) a strong s-process occurs, powered by the reaction chain $^{12}C(p,\gamma)\,^{13}N(\beta^+)\,^{13}C(\alpha,n)\,^{16}O$ where the protons and α-particles are provided by the $^{12}C + {}^{12}C$ fusion reaction. Besides possible nucleosynthesis interest, this reaction chain is important for the evolution of the neutron excess η and therefore necessary for the description of the global parameters of core evolution.

The results presented here are to be compared with hydrostatic nucleosynthesis from stellar evolution calculations; the most complex nuclear treatment was performed by Weaver, Zimmerman, and Woosley (1978) and Weaver, Woosley, and Fuller (1983). In both cases a reaction network containing 19 nuclear species was used up to O-burning. Then Weaver, Zimmerman, and Woosley (1978) switched to a complete nuclear statistical equilibrium whereas Weaver, Woosley, and Fuller (1983) used a treatment where two QSE networks around Si and Fe are linked together by the reaction $^{45}Sc(p,\gamma)\,^{46}Ti$. This change, together with improved electron capture rates (Fuller, Fowler, and Newman 1982), led to a smaller Fe-core mass (after Si-burning) of 1.37 $M_\odot$ for a 25 $M_\odot$ star compared to a 1.6 $M_\odot$ Fe-core in the past calculation. That change shows the drastic dependence on nuclear input parameters. The smaller mass of the core is in better agreement with Arnett (1977), in which a simpler treatment of charged particle reactions was balanced by a more careful treatment of electron capture.

It has been recently pointed out (Hillebrandt 1982; Arnett 1982; Hillebrandt 1984) how important the mass of the Fe-core is to a successful type II supernova explosion. This motivates this investigation with a 254 nuclei reaction network of how complex minimum size networks have to be in order to assure the necessary accuracy of η and Y_e.

After discussing the successive burning stages from Ne to Si burning in §§ 11.6 and 11.7, § 11.8 is devoted to the evolution of global parameters. It is shown that for "relatively" low mass stars ($M_\alpha < 4\ M_\odot$, $M < 15\ M_\odot$) electron capture in O-burning leads to a strong neutronization, increasing η and resulting in ^{30}Si being the most abundant nucleus in early phases of Si-burning rather than ^{28}Si. In those cases the link between the two QSE groups is not given by ^{45}Sc(p,γ) ^{46}Ti but the reaction flow is rather passing through the neutron-rich Ca-isotopes.

Large uncertainties (besides the treatment of convection) still exist in the final evolution of massive stars, and better calculations have still to be done.

In § 11.9 we try to perform a preliminary application to hydrostatic shell burning by (mis)using the code developed for polytropic core burning to simulate typical shell burning conditions. When adding up those shell burning products weighted with the appropriate shell masses according to earlier evolution calculations (Arnett 1972–1977; Woosley and Weaver 1982), we find a qualitative agreement with type II supernova abundance yields of a 25 $M_\odot$ star (Woosley and Weaver 1982). This includes an overproduction of Ne, Na, and Mg while Ar, Ca, Ti, etc., are underproduced. The first group of elements originates from C-burning while the latter is produced in high temperature O-burning. When the ^{12}C(α,γ) ^{16}O rate is varied within the present uncertainties (up to a factor of 4 over the rate given by Fowler, Caughlan, and Zimmerman 1975; see Kettner et al. 1982 and Langanke and Koonin 1983), these features can be removed, giving a rather constant overabundance curve for the elements from O to Ge in the pre-explosive composition. This indicates that an average type II supernova could produce those elements in relative solar proportions.

This seems to solve the present dilemma of explaining solar abundances with nucleosynthesis products of type II supernovae; a consistency check must be done by performing a realistic stellar evolution calculation as well as testing the influence of explosive processing on the precollapse abundances of prior hydrostatic burning.

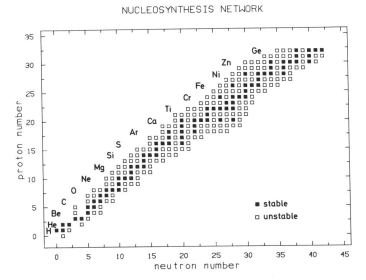

Fig. 11.1. Nucleosynthesis network employed in the hydrostatic nucleosynthesis calculations, containing 254 nuclear species from H to ^{74}Ge. Filled squares denote stable, open squares unstable nuclei. The inclusion of neutron-rich species allows examination of the onset of the *s*-process.

11.2 Nucleosynthesis Network

The most complex stellar *evolution* code (KEPLER) in respect to nucleosynthesis details (Weaver, Zimmerman, and Woosley 1978) contains 19 nuclear species for all burning stages prior to silicon burning and then changes to an equilibrium network. The simplified structure assumptions discussed in the following section allow for an extensive treatment of detailed nucleosynthesis to explore all possible effects, neglected or not observed in earlier calculations. The network employed in this work is shown in figure 11.1, containing $N = 254$ nuclear species from neutrons and protons to ^{74}Ge and suited for essentially all burning stages. This is larger than earlier networks used in explosive nucleosynthesis (Woosley, Arnett, and Clayton 1973; $N = 84$) and also larger than the extensive network applied by Woosley and Weaver (1982) ($N = 131$, up to Ni) for the postprocessing of all zones from the earlier evolution (Weaver et al. 1978) and explosion calculations (Weaver and Woosley

1980; Woosley and Weaver 1982) which followed the known temperature and density histories. Besides other differences, the network applied here extends to more neutron-rich nuclei and allows examination of the influence of neutron exposures during different burning stages.

The nuclear reaction rates are, whenever possible, taken from Fowler, Caughlan, and Zimmerman (1975) including the update by Fowler (1981). When, in this compilation, rate expressions involve uncertainty factors θ with values in the range [0,1], the adopted standard value is always $\theta = 0.1$. For heavier nuclei involving a higher density of compound nuclear resonances the statistical model calculations from Woosley et al. (1975, 1978) are included and complemented in some cases by results obtained with the statistical model code SMOKER (Thielemann 1980). The new and/or corrected (mostly) proton capture rates on proton-rich nuclei by Wallace and Woosley (1981) are taken into account; when necessary the rough estimates of Wagoner et al. (1967) and Wagoner (1969) are used. Weak interaction rates applied here are those from Fuller, Fowler, and Newman (1980, 1982) for the β-decay and electron capture for 166 nuclei. When nuclei are not contained in this compilation, experimental (terrestrial) half-lives (Seelman-Eggebert et al. 1981) are used.

The system of ordinary differential equations

$$\dot{Y_i} = \sum_j c_i(j)\lambda_j \; Y_j + \sum_{j,k} c_i(j,k)\rho \; N_A <jk> \; Y_j \; Y_k$$
$$+ \sum_{j,k,l} c_i \, (j,k,l) \; \rho^2 \; N_A^2 \; <jkl> \; Y_j \; Y_k \; Y_l \tag{1}$$

for the nuclear abundances $Y_i = n_i/\rho N_A$ (see, e.g., Fowler et al. 1975, n_i number density, ρ mass density, N_A Avogadro's number) has been solved numerically with the Euler-backward-differentiation method for stiff systems (Gear 1971; Arnett and Truran 1969).

11.3 Stellar Model

11.3.1 Method

In this investigation we will concentrate on the nucleosynthesis occurring in convective cores of stars. The evolution equations are reduced from partial differential equations to ordinary differential equations by performing an integration over the stellar structure; this requires certain (not unreasonable)

assumptions to be made. The general ideas for this approach are outlined by
Fowler and Hoyle (1964) and Arnett (1978).

We assume the stars to be in hydrostatic equilibrium (except for convective mixing, which will occur for all burning stages we consider). The structure of the cores may be approximated by a polytrope of index n where
$1.5 \leq n \leq 3$. The central pressure p_c, density ρ_c, and core mass M are then
related by (Chandrasekhar 1939)

$$p_c = D_n \, G \, M^{2/3} \, \rho_c^{4/3} \tag{2}$$

where G is the gravitational constant and D_n is a slowly varying function of
polytropic index, being 0.364 for $n = 3$ and 0.478 for $n = 1.5$. To be definite
we simply take $n = 3$ in what follows. The first law of thermodynamics is

$$\dot{E} + p\dot{V} = \epsilon - \frac{dL}{dM} - S_\nu - S_{\nu\beta} \tag{3}$$

where ϵ is the energy generation rate, dL/dM the energy loss rate due to radiative flux, S_ν the loss rate due to different neutrino emission processes (pair
annihilation neutrinos, photo neutrinos, and plasmon neutrinos), and $S_{\nu\beta}$ the
loss rate due to ν-emission accompanying β-decay or electron capture. In case
of a polytropic core the integration over the stellar structure leads to

$$\dot{E}_c + p_c \dot{V}_c = \begin{Bmatrix} 1.608 \\ 1.885 \end{Bmatrix} \times \left[\epsilon_c \frac{<\epsilon>}{\epsilon_c} - \frac{L}{M} - S_{\nu,c} \frac{<S_\nu>}{S_{\nu,c}} \right.$$

$$\left. - S_{\nu\beta,c} \frac{<S_{\nu\beta}>}{S_{\nu\beta,c}} \right], \tag{4}$$

where the different numbers correspond to a polytropic index of $n = 3$ or 1.5,
respectively. This result (see Arnett and Thielemann 1984 for the derivation)
enables us to describe the evolution of the core in terms of its central values.
When the temperature and density dependence of the energy generation and
loss terms can be approximated by $\epsilon = \epsilon_c \, (\rho/\rho_c)^{u-1} \, (T/T_c)^s$, the effective factors $<\epsilon>/\epsilon_c$ of averaged to central values can be easily obtained. Fowler
and Hoyle (1964) have shown that if functions like ϵ, S_ν, and $S_{\nu\beta}$ are sharply
peaked near the center due to their temperature dependence, then

$$\frac{<\epsilon>}{\epsilon_c} = \begin{Bmatrix} 3.23 \\ 2.40 \end{Bmatrix} (nu + s)^{-3/2} \ . \tag{5}$$

The same holds true for S_ν and $S_{\nu\beta}$. Together with an equation of state

$$p = p(\rho, T, Y_i\,; \quad i{=}1,...N) \text{ and } E = E(\rho, T, Y_i\,; \quad i{=}1,...N) \tag{6}$$

which yields the pressure and internal energy as a function of temperature, density, and chemical composition and the set of N equations (1) for the chemical composition, equations (2) and (4) can be used to determine the basic thermodynamical quantities ρ and T and follow the evolution of the stellar core. The equation of state used in this computation (Arnett 1969) includes an ideal ion gas, radiation pressure, and the noninteracting Fermi-Dirac gas of electrons and positrons. The right-hand side of equation (4) has several different energy gain and loss terms. What is conventionally called the "energy generation rate" (the energy released by nuclear reactions) is denoted by

$$\epsilon = - \sum_i \dot{Y}_i\, N_A\, \Delta M_i\ , \tag{7}$$

where N_A is Avogadro's number, $\Delta M_i = (M_i - A_i M_u)c^2$ is the mass excess in energy units (Wapstra and Bos 1977), A_i is the mass number, and $M_u = 1/N_A$ the atomic mass unit.

During the early stages of hydrogen and helium burning, the important energy loss term is L/M, which means the star is cooled by radiative diffusion (outside the convective core). We consider fairly massive stars, so that the opacity is due to Thomson scattering, and $\kappa \approx 0.4\ Y_e$, where Y_e is the number of electrons per baryon defined like Y_i in equation (1). Using Eddington's standard model, we have

$$\frac{L}{M} = 1.88 \times 10^2\, \frac{1}{Y_e} \left[\frac{M}{M_\odot} \right]^2 \left[\frac{\beta}{Y} \right]^4 \text{ erg/gs}\ ,$$

$$\text{with}\ \ Y = Y_e + \sum_i Y_i\ , \tag{8}$$

and β is the ratio of gas pressure to total pressure. Strictly speaking, this is not exact because β is not constant throughout the star and the gas is not ideal. However, for massive stars during helium burning the approximation is quite good.

The neutrino energy loss rate S_ν is the most important cooling mechanism from carbon burning onward, and may be found to adequate accuracy by standard expressions of Beaudet, Petrosian, and Salpeter (1967), slightly scaled to approximate the effects of neutral currents (Ramadurai 1976).

The neutrino losses accompanying β-decay and electron capture $S_{\nu\beta}$ are important in oxygen and silicon burning, and were taken from the extensive compilation of Fuller, Fowler, and Newman (1980, 1982) as were λ_β and λ_{e^-} used in equation (1).

The coefficients u and s needed for the factors $<\epsilon>/\epsilon_c$, $<S_\nu>/S_{\nu,c}$, and $<S_{\nu\beta}>/S_{\nu\beta,c}$ can be obtained by fitting the energy generation and loss terms to a $(\rho/\rho_0)^{u-1}\,(T/T_0)^s$ form. This is easy for the neutrino losses which are given analytically in ρ and T, and is determined from the numerical solution of the nucleosynthesis network for ϵ and $S_{\nu\beta}$.

This, as well as the numerical technique to solve the nucleosynthesis equations (1), the energy equation (4), and the equation of hydrostatic equilibrium (2) simultaneously, is given in more detail in Arnett and Thielemann (1984).

With this system of equations we can march through the evolution of a given stellar core, updating T, ρ, and the set Y_i over suitably chosen time steps. This is a relatively minor modification to a reaction network code; some logic, the neutrino loss rates, and a realistic equation of state are the major additions. We have reduced the computational effort by a factor of order n, where n is the number of zones considered in a direct solution.

11.3.2 Core Evolution

The method described in § 11.3.1 has been applied for He-cores of 1.5, 2, 4, 8, and 16 $M_\odot$. The initial abundances were solar system values (Cameron 1982), modified by CNO burning of hydrogen (Arnould and Nørgaard 1978.)

A strictly self-consistent treatment of nucleosynthesis and core evolution in a simple model is not possible. Calculations of this kind can be used to explore detailed nucleosynthesis effects during all burning stages, but must be compared to temperature and density conditions from evolutionary calculations. Then an "effective polytropic core" mass, M_{eff}, can be used in the model. This mass will be a fairly subtle function of evolutionary state, being different from both the helium core mass and the convective core mass for the late burning stages. We chose the pragmatic approach of finding a value of $M_{\text{eff},i}$ such that a polytropic core reproduces in a reasonable way the central evolution of each subsequent burning stage i ($i =$ He, C, Ne, O, Si). Switching from one burning stage to the next means a reduction in M_{eff} by

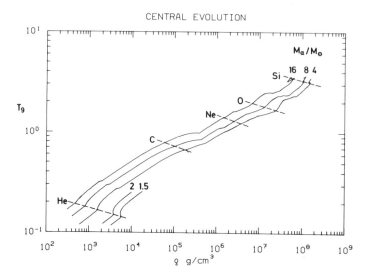

Fig. 11.2. For the 4, 8, and 16 $M_\odot$ He-core (corresponding approximately to 15, 25, and 45 $M_\odot$ stars) the evolution is followed through C, Ne, O, and Si ignition. These subsequent burning stages have been carried through to simulate burning conditions of stellar evolution calculations (Arnett 1972–1977) by reducing the mass of the polytropic core, keeping the central pressure constant during this adjustment (see eq. [12]).

$\Delta M_{\text{eff}} = M_{\text{eff},i} - M_{\text{eff},\,i+1}$ which was spread over 100 time steps under the assumption of constant central pressure during the adjustment, solving the equations

$$p_c = D_n \, G \, M_{\text{eff}}{}^{2/3} \, \rho_c^{4/3} = \text{const}$$

$$M_{\text{eff}}(t + \Delta t) = M_{\text{eff}}(t) - \Delta M_{\text{eff}}/100 \qquad (9)$$

in between two evolutionary time steps. With the additional aid of the equation of state, $p = p(\rho, T)$, the change in temperature is found. The result of this procedure is shown in figure 11.2 for 1.5, 2, 4, 8, and 16 $M_\odot$ He-cores. While the exact details (including all wiggles and loops) of the evolutionary calculations (Arnett 1972a, b, 1973, 1974a, b, 1977) cannot be simulated, the general trend is well reproduced. This also holds true for the appearance of s-shaped crooks at the onset of some burning stages as well as for the silicon flash(es).

The onset of burning stages is marked by dashed lines. Carbon burning had to be treated slightly differently. This is not surprising, as neutrino cooling becomes dominant in this phase. To come close to the conditions of Arnett (1972b) the mass adjustment of M_{eff} from He- to C-burning had to be spread over 300 time steps. Endal (1975) and Weaver, Zimmerman, and Woosley (1978) find similar conditions for central carbon burning. The evolutionary tracks so obtained will be taken as a basis for discussing the nucleosynthesis in subsequent core burning stages.

11.4 Core Helium Burning

11.4.1 General Nucleosynthesis

The abundance pattern from helium burning is of special interest since the recent changes in the $^{12}C(\alpha,\gamma)$ ^{16}O reaction cross section (Kettner et al. 1982). The reported modification (a factor of 2 to 5 compared with Fowler et al. 1975) will change the $^{12}C/^{16}O$ ratio at the end of He-burning and thus change the fuel for later burning stages, which will have a dominant influence on the final pre-supernova abundances. Fowler (this conference) now quotes a factor of 2–4 based on the reanalysis of the Caltech and Münster data by Langanke and Koonin (1983).

The use of a large network up to ^{74}Ge allows a detailed description of the lighter elements as well as the onset of the s-process, enabling a comparison to Couch et al. (1974) and Lamb et al. (1977). The exact factor by which the $^{12}C(\alpha,\gamma)$ ^{16}O reaction is changed is not yet clear, as it still involves experimental uncertainties and a theoretical extrapolation from experimental to astrophysically low reaction energies. We performed calculations for three cases with the factors $f = 1$, 3, and 5.

The results for the main ashes of helium burning (^{12}C and ^{16}O) are given in table 11.1 for a sequence of He-core masses and the three $^{12}C(\alpha,\gamma)$ ^{16}O rates mentioned above. As the evolutionary calculations by Arnett (1972a) and the s-process calculations for core He-burning by Couch et al. (1974) made use of a $^{12}C(\alpha,\gamma)$ ^{16}O reaction given in Fowler et al. (1967) which differs by a factor f between 1.56 and 1.78 in the range $0.1 \leq T_9 \leq 0.2$, we expect their results to lie in between cases 1 and 3. This is in fact the case: Arnett (1972a) reports $X_c = 0.434$ for $M_\alpha = 4\,M_\odot$ and $X_c = 0.358$ for $M_\alpha = 8\,M_\odot$. Table 11.1 shows the drastic change of the C/O ratio after core He-burning with

Table 11.1
Carbon and Oxygen Mass Fraction after Helium Burning

f	1		3		5	
$M_\alpha/M_\odot$	^{12}C	^{16}O	^{12}C	^{16}O	^{12}C	^{16}O
1.5	0.720	0.248	0.524	0.443	0.403	0.565
2	0.696	0.272	0.488	0.480	0.364	0.605
4	0.625	0.341	0.396	0.571	0.264	0.704
8	0.547	0.416	0.298	0.669	0.164	0.802
16	0.466	0.494	0.202	0.760	0.075	0.883

f = multiplication factor used for the $^{12}C(\alpha,\gamma)$ ^{16}O reaction of Fowler et al. (1975).

increased $^{12}C(\alpha,\gamma)$ ^{16}O rates. When taking $M_\alpha = 8\ M_\odot$, which corresponds to a 20–25 $M_\odot$ star and therefore a typical type II supernova progenitor, the carbon mass fraction goes down from 55% to 16% and the oxygen goes up from 42% to 80%, which is of great importance for the later evolution during carbon and oxygen burning and pre-supernova abundances. It should be noted that both stellar evolution calculations by Ober (private communication) and our findings of slightly higher temperatures in He-burning in Arnett (1972a) suggest even less ^{12}C and more ^{16}O than quoted in table 11.1.

The results for specific nuclei up to ^{58}Fe which are either involved in the main He-burning reactions $(3\alpha \rightarrow\ ^{12}C(\alpha,\gamma)\ ^{16}O(\alpha,\gamma)\ ^{20}Ne$ and $^{14}N(\alpha,\gamma)\ ^{18}F$, $^{18}F(\beta^+)\ ^{18}O(\alpha,\gamma)\ ^{22}Ne(\alpha,n)\ ^{25}Mg(n,\gamma)\ ^{26}Mg)$ or highly overproduced by neutron s-processing are displayed in table 11.3 for the case of the 8 $M_\odot$ He-core. A list of important reactions during He-burning down to a factor of 100 of major energy generation reactions is given in table 11.2. Comparing the result of table 11.3 with Lamb et al. (1977) and Couch et al. (1974) gives a consistent picture when taking into account the previous comments. Column 1 $(f = 1)$ of table 11.3 $(M_\alpha = 8\ M_\odot)$ is comparable with the results for $M = 25\ M_\odot$ by Lamb et al. (1977). Taking the appropriate value for the rate of $^{22}Ne(\alpha,n)\ ^{25}Mg$, the abundances given by Couch et al. (1973) correspond to the respective He-core masses of this work when values intermediate between columns 1 and 2 are used (this is because their $^{12}C(\alpha,\gamma)\ ^{16}O$ rate corresponds to $f \approx 1.65$). Slight changes are due to minor modifications in solar abundances (Cameron 1973 to Cameron 1982) and improvements in thermonuclear reaction rates (Woosley et al. 1978). The conclusions are similar: "light nuclei" produced significantly overabundant over solar in stars more massive

Table 11.2
Important Reactions in He-Burning

(a) energy generation

$$^4\text{He}(2\alpha,\gamma)\ ^{12}\text{C}(\alpha,\gamma)\ ^{16}\text{O}(\alpha,\gamma)\ ^{20}\text{Ne}$$

(b) neutron source

$$^{14}\text{N}(\alpha,\gamma)\ ^{18}\text{F}(\beta^+)^{18}\text{O}(\alpha,\gamma)\ ^{22}\text{Ne}\ (\alpha,\gamma)\ ^{26}\text{Mg}$$
$$^{14}\text{N}(\alpha,\gamma)\ ^{18}\text{F}(\beta^+)^{18}\text{O}(\alpha,\gamma)\ ^{22}\text{Ne}\ (\alpha,n)\ ^{25}\text{Mg}$$

(c) high temperature burning with effective $^{22}\text{Ne}(\alpha,n)\ ^{25}\text{Mg}$

$$^{22}\text{Ne}(n,\gamma)\ ^{23}\text{Ne}(\beta^-)\ ^{23}\text{Na}(n,\gamma)\ ^{24}\text{Na}(\beta^-)\ ^{24}\text{Mg}$$
$$^{20}\text{Ne}(n,\gamma)\ ^{21}\text{Ne}(\alpha,n)\ ^{24}\text{Mg}$$
$$^{24}\text{Mg}(n,\gamma)\ ^{25}\text{Mg}(n,\gamma)\ ^{26}\text{Mg}(n,\gamma)\ ^{27}\text{Mg}(\beta^-)\ ^{27}\text{Al}$$
$$^{27}\text{Al}(n,\gamma)\ ^{28}\text{Al}(\beta^-)\ ^{28}\text{Si}(n,\gamma)\ ^{29}\text{Si} + \text{further s-processing}$$

---

than 15 $M_\odot$ are ^{21}Ne, ^{22}Ne, ^{25}Mg, ^{26}Mg, ^{36}S, ^{58}Fe, and partially ^{37}Cl. The case of ^{40}Ar is somewhat confused as the solar abundance changed drastically. In both cases it is a theoretical estimate. Anders and Ebihara (1982) quote Göbel et al. (1978), who found the lowest meteoritic $^{40}\text{Ar}/^{36}\text{Ar}$ ratio to be $(2.9 \pm 1.7) \times 10^{-4}$. When assuming that the dominant part of ^{40}Ar in the solar system is not due to ^{40}K decay, which is the case if both nuclei are produced with equal abundances because only 11% of ^{40}K decays to ^{40}Ar, this would lead to a solar abundance of 25.4 (Si $= 10^6$) instead of 200 (Cameron 1982) and the $Y/Y_\odot$ ratios given in table 11.3 would have to be multiplied by 7.87. That would lead to a similar enhancement of ^{40}Ar as found for $^{21,22}\text{Ne}$, $^{25,26}\text{Mg}$, ^{36}S, and ^{58}Fe. This may be the source of ^{40}Ar.

What can also be seen from table 11.3 is the change in s-processing when $^{12}\text{C}(\alpha,\gamma)\ ^{16}\text{O}$ is increased by a factor f of 3 (column 2) to 5 (column 3). An enhanced $^{12}\text{C}(\alpha,\gamma)\ ^{16}\text{O}$ rate reduces the α-particles available for the $^{22}\text{Ne}(\alpha,n)\ ^{25}\text{Mg}$ reaction which is reponsible for the neutron production. This effect can decrease the total neutron production by a factor of 2 (in column 3) and gives the corresponding reduction in overabundances of $^{25,26}\text{Mg}$, ^{36}S, ^{40}Ar, and s-process nuclei. This is also seen in figure 11.3, which shows the

Table 11.3
$M_\alpha = 8\,M_\odot$

f	1		3		5	
Isotope	Y	$Y/Y_\odot$	Y	$Y/Y_\odot$	Y	$Y/Y_\odot$
^{12}C	4.56×10^{-2}	$1.42 \times 10^{+2}$	2.48×10^{-2}	$7.69 \times 10^{+1}$	1.37×10^{-2}	$4.21 \times 10^{+1}$
^{16}O	2.60×10^{-2}	$4.87 \times 10^{+1}$	4.18×10^{-2}	$7.82 \times 10^{+1}$	5.01×10^{-2}	$9.38 \times 10^{+1}$
^{20}Ne	1.18×10^{-4}	1.77	1.07×10^{-4}	1.59	1.15×10^{-4}	1.71
^{21}Ne	7.51×10^{-6}	$3.70 \times 10^{+1}$	8.32×10^{-6}	$4.09 \times 10^{+1}$	8.64×10^{-6}	$4.25 \times 10^{+1}$
^{22}Ne	4.41×10^{-4}	$5.39 \times 10^{+1}$	6.94×10^{-4}	$8.47 \times 10^{+1}$	7.16×10^{-4}	$8.74 \times 10^{+1}$
^{24}Mg	4.66×10^{-5}	1.93	3.60×10^{-5}	1.49	3.53×10^{-5}	1.46
^{25}Mg	3.49×10^{-4}	$1.12 \times 10^{+2}$	2.41×10^{-4}	$7.77 \times 10^{+1}$	2.33×10^{-4}	$7.50 \times 10^{+1}$
^{26}Mg	3.05×10^{-4}	$8.91 \times 10^{+2}$	1.92×10^{-4}	$5.60 \times 10^{+1}$	1.81×10^{-4}	$5.27 \times 10^{+1}$
^{36}S	3.62×10^{-7}	$1.84 \times 10^{+2}$	2.48×10^{-7}	$1.25 \times 10^{+2}$	2.40×10^{-7}	$1.22 \times 10^{+2}$
^{37}Cl	1.45×10^{-6}	$4.30 \times 10^{+1}$	1.35×10^{-6}	$4.01 \times 10^{+1}$	1.34×10^{-6}	$3.97 \times 10^{+1}$
^{40}Ar	7.41×10^{-8}	$1.32 \times 10^{+1}$	4.85×10^{-8}	8.73	4.68×10^{-8}	8.43
^{40}K	2.55×10^{-8}		2.03×10^{-8}		1.96×10^{-8}	
^{58}Fe	8.52×10^{-6}	$9.89 \times 10^{+1}$	1.07×10^{-5}	$1.24 \times 10^{+2}$	1.08×10^{-5}	$1.25 \times 10^{+2}$

corresponding onset of the s-process for nuclei up to $A = 74$. Higher temperatures for He-burning in the most massive case ($M_\alpha = 16\ M_\odot$) result in the strongest s-process. With an increased $^{22}\text{Ne}(n,\gamma)\ ^{23}\text{Ne}$ rate as reported by Almeida and Käppeler (1983), which is still consistent with the value given by Woosley et al. (1975, 1978) within the quoted error bars, ^{22}Ne becomes a larger neutron poison and the strength of the s-process is also weakened.

11.5 Core Carbon Burning

The most important reaction is $^{12}\text{C}(^{12}\text{C},\alpha)\ ^{20}\text{Ne}$ and $^{12}\text{C}(^{12}\text{C},p)\ ^{23}\text{Na}$ with almost equal branching. Most of the ^{23}Na is transformed into ^{20}Ne by $^{23}\text{Na}(p,\alpha)\ ^{20}\text{Ne}$, ranging between 60 and 75% depending on the temperature and density involved. The second highest destruction rate of ^{23}Na is $^{23}\text{Na}(p,\gamma)\ ^{24}\text{Mg}$. This means that the most abundant product is ^{20}Ne followed by ^{23}Na and ^{24}Mg. In addition, ^{16}O has a large abundance after He-burning. Other important reactions, with intensities less by roughly one order of magnitude, are

$$^{20}\text{Ne}(\alpha,\gamma)\ ^{24}\text{Mg}, \qquad\qquad ^{23}\text{Na}(\alpha,p)\ ^{26}\text{Mg}(p,\gamma)\ ^{27}\text{Al},$$
$$^{20}\text{Ne}(n,\gamma)\ ^{21}\text{Ne}(\alpha,n)\ ^{24}\text{Mg}, \qquad ^{21}\text{Ne}(p,\gamma)\ ^{22}\text{Na}(\beta^+)\ ^{22}\text{Ne}(\alpha,n)\ ^{25}\text{Mg}(n,\gamma)\ ^{26}\text{Mg}$$

accompanied by

$$^{22}\text{Ne}(p,\gamma)\ ^{23}\text{Na} \quad\text{and}\quad ^{25}\text{Mg}(p,\gamma)\ ^{26}\text{Al}(\beta^+)\ ^{26}\text{Mg},$$

with ^{25}Mg being the most efficient neutron poison. For stars with original He-cores less than 8 $M_\odot$,

$$^{12}\text{C}(p,\gamma)\ ^{13}\text{N}(\beta^+)\ ^{13}\text{C}(\alpha,n)\ ^{16}\text{O}(\alpha,\gamma)\ ^{20}\text{Ne}$$

is also important and the main source of neutrons. Those stars burn carbon at low temperatures and higher densities, both favor the production of ^{13}N via $^{12}\text{C}(p,\gamma)\ ^{13}\text{N}$ while for higher temperatures ($T_9 > 0.8$) this reaction is counterbalanced by the photodisintegration via $^{13}\text{N}(\gamma,p)\ ^{12}\text{C}$. This reaction has a low Q-value of 1.94 MeV and thus needs only low temperature photons for the inverse reaction. Arnett and Truran (1969) mentioned this neutron source in C-burning. Lower mass stars burn He at lower temperatures and leave more ^{12}C fuel for the above reaction instead of transforming it into ^{16}O. For low temperature carbon burning, $^{13}\text{C}(\alpha,n)\ ^{16}\text{O}$ acts as a strong neutron source, becoming more important than $^{22}\text{Ne}(\alpha,n)\ ^{26}\text{Mg}$ in central He-burning and

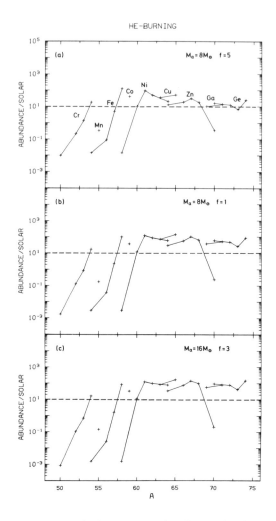

Fig. 11.3. *s*-processing during core He-burning. The low mass He-cores ($M_\alpha/M_\odot = 1.5$, 2) show almost no enhancement of *s*-process nuclei over solar (abundance over solar $\sim 10^0$) except for ^{58}Fe. For more massive stars, burning He at higher temperatures, *s*-processing becomes stronger because of a faster ^{22}Ne(α,n) ^{25}Mg reaction. The variation in the ^{12}C(α,γ) ^{16}O rate tends to be less influential for *s*-processing, and in the case of $M_\alpha = 16$ $M_\odot$ an almost identical neutron flux is obtained (note the ^{70}Zn is not an *s*-process nucleus).

causing strong s-processing. Under those conditions the sequence

$$^{21}\text{Ne}(n,\gamma)\ ^{22}\text{Ne}(n,\gamma)\ ^{23}\text{Ne}(\beta^-)\ ^{23}\text{Na}(n,\gamma)\ ^{24}\text{Na}(\beta^-)\ ^{24}\text{Mg}$$

also becomes important. The reactions $^{24}\text{Mg}(p,\gamma)\ ^{25}\text{Al}(\beta^+)\ ^{25}\text{Mg}$ occur under the same circumstances, because the first reaction has a small Q-value of 2.27 MeV so that ^{25}Al is stable against photodisintegration only for low temperatures. The reactions discussed above are a subset of the list Howard et al. (1972) give as important reactions for *explosive* carbon burning. The quantities $S_n = \int X_n\, \rho\, dt$ or $\tau_{30\,\text{keV}} = v_T \int n(t)\, dt$ (kT = 30 keV) describe the strength of the s-process discussed above. The case $f = 1$ leads to the strongest s-processing because the largest amount of ^{12}C is left from He-burning; this eventually produces ^{13}C via $^{12}\text{C}(p,\gamma)\ ^{13}\text{N}(\beta^+)\ ^{13}\text{C}$.

For the 8 $M_\odot$ He-core (which might be a type II supernova progenitor typical for nucleosynthesis), the neutron exposure is very similar in He-burning and in C-burning. As the material experienced both exposures sequentially, the result is a combined effect, $\tau = \tau_{\text{He}} + \tau_{\text{C}}$, where $\tau = 0.331$ for the 8 $M_\odot$ He-core. This is near to but larger than the $\tau = 0.24$ of Käppeler et al. (1982) for an exponential distribution of exposures to reproduce the heavy s-nuclei up to Bi. New experimental results of neutron capture cross sections for the Nd-isotopes (Mathews and Käppeler 1984) indicate a value close to 0.3, however. We see an almost constant overabundance in this case for s-nuclei above $A = 60$ (figure 11.4b), and the jump at ^{74}Ge is due to the limited network. Figures 11.4a and 11.4c show the dependence on the mass of the original He-core. In the 4 $M_\odot$ He-core s-processing is much stronger. This nucleosynthesis can be important for the further evolution, as can the increase of the neutron excess η of the core via the $^{13}\text{N}(\beta^+)\ ^{13}\text{C}$ reaction. It is, however, not clear if this material will ever by ejected in this form to contribute to nucleosynthesis yields. From carbon burning on, central cores are of the size of one Chandrasekhar mass (M_{Ch}). This material will be transformed further by the following burning stages and locked up in the neutron star probably formed by a type II supernova. The carbon burning matter being ejected has only experienced higher (shell) burning temperatures where the $^{12}\text{C}(p,\gamma)\ ^{13}\text{N}(\beta^+)\ ^{13}\text{C}(\alpha,n)\ ^{16}\text{O}$ mechanism will not be effective. A summary of important reactions is given in table 11.4.

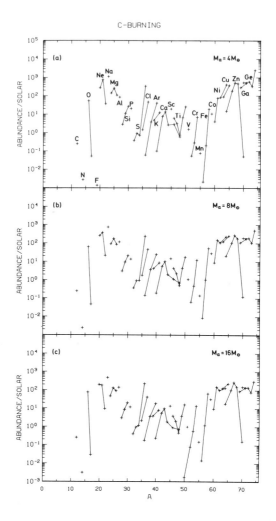

Fig. 11.4. Abundances after core C-burning. Generally a strong enhancement in Ne, Na, and Mg is obtained. In addition, the combined influence of neutron captures from prior He-burning and C-burning is seen in s-process isotopes above Fe as well as neutron-rich isotopes of medium elements (e.g., ^{29,30}Si, ^{36}S, ^{37}Cl, ^{40}Ar, etc.). Note strong s-process for the 4 $M_\odot$ He-core due to the ^{13}C(α,n) ^{16}O reaction in C-burning.

11.6 Core Neon and Oxygen Burning

The important reactions in Ne-burning are summarized in table 11.5, but here we concentrate on the more interesting phase of O-burning. Explosive

Table 11.4
Important Reactions in Carbon Burning

(a) basic reactions

$$^{12}C\left(^{12}C, ^{\alpha}_{p}\right) \begin{array}{l} ^{20}Ne \\ ^{23}Na \end{array}$$

$^{23}Na(p,\alpha)\, ^{20}Ne \qquad ^{23}Na(p,\gamma)\, ^{24}Mg$

(b) down to 10^{-2} of (a)

$^{20}Ne(\alpha,\gamma)\, ^{24}Mg \qquad ^{23}Na(\alpha,p)\, ^{26}Mg(p,\gamma)\, ^{27}Al$

$^{20}Ne(n,\gamma)\, ^{21}Ne(p,\gamma)\, ^{22}Na(\beta^+)\, ^{22}Ne(\alpha,n)\, ^{25}Mg(n,\gamma)\, ^{26}Mg$

$\qquad\qquad (\alpha,n)\, ^{24}Mg \qquad (p,\gamma)\, ^{23}Na \quad (p,\gamma)\, ^{26}Al(\beta^+)\, ^{26}Mg$

(c) low temperature, high density

$^{12}C(p,\gamma)\, ^{13}N(\beta^+)\, ^{13}C(\alpha,n)\, ^{16}O(\alpha,\gamma)\, ^{20}Ne$

$^{24}Mg(p,\gamma)\, ^{25}Al(\beta^+)\, ^{25}Mg$

$^{21}Ne(n,\gamma)\, ^{22}Ne(n,\gamma)\, ^{23}Ne(\beta^-)\, ^{23}Na(n,\gamma)\, ^{24}Na(\beta^-)\, ^{24}Mg + s\text{–processing}$

Table 11.5
Neon Burning

(a) basic reactions

$^{20}Ne(\gamma,\alpha)\, ^{16}O \qquad\qquad ^{20}Ne(\alpha,\gamma)\, ^{24}Mg(\alpha,\gamma)\, ^{28}Si$

(b) flows $> 10^{-2} \times$ (a)

$^{23}Na(p,\alpha)\, ^{20}Ne \qquad\qquad ^{23}Na(\alpha,p)\, ^{26}Mg(\alpha,n)\, ^{29}Si$

$^{20}Ne(n,\gamma)\, ^{21}Ne(\alpha,n)\, ^{24}Mg(n,\gamma)\, ^{25}Mg(\alpha,n)\, ^{28}Si$

$^{28}Si(n,\gamma)\, ^{29}Si(n,\gamma)\, ^{30}Si$

$^{24}Mg(\alpha,p)\, ^{27}Al(\alpha,p)\, ^{30}Si$

$^{26}Mg(p,\gamma)\, ^{27}Al(n,\gamma)\, ^{28}Al(\beta^-)\, ^{28}Si$

(c) at low temperature and high density
(^{22}Ne left from prior n-rich C-burning)

$^{22}Ne(\alpha,n)\, ^{25}Mg(n,\gamma)\, ^{26}Mg(n,\gamma)\, ^{27}Mg(\beta^-)\, ^{27}Al$

---

oxygen burning has been discussed in detail by Truran and Arnett (1970) and Woosley, Arnett, and Clayton (1973). Under explosive conditions it is in some cases hard to distinguish between explosive O- and Si-burning. Hydrostatic burning leads to fewer important reactions, which are displayed in table 11.6. Categories given are (a) basic reactions in the sense of those occurring in all calculations done for different masses, (b) additional reactions occurring in higher temperature burning of the 16 $M_\odot$ He-core, and (c) additional reactions being of importance in lower temperature, higher density burning of the original 4 $M_\odot$ He-core. The general trend is that the higher temperature burning allows the buildup of heavier nuclei by overcoming a higher Coulomb barrier, while low temperature, high density burning leaves a more degenerate electron gas and enhances, generally speaking, electron captures on nuclei which again enhances the neutron excess.

For a 16 $M_\odot$ He-core, η increases during O-burning from 2.3×10^{-3} to 10^{-2} (roughly a factor of 4) while the factor is close to 8 for the 4 $M_\odot$ core. This enhanced electron capture also leads to differences in energy losses via neutrinos, which will be discussed later.

The nucleosynthesis results for central O-burning, dependent on the initial He-core mass, are given in figure 11.5. The main burning products are ^{28,30}Si, ^{34}S, and ^{38}Ar. The more massive stars also produce ^{42}Ca and ^{46}Ti. While these latter nuclei form about or less than 1% of the total mass, the steep decrease of solar abundances beyond ^{28}Si by four orders of magnitude leads to a similar production over solar as for the main products like ^{28}Si.

During O-burning, temperatures favor the photodisintegration of heavy nuclei (produced by the s-process) into Fe-peak nuclei. This is seen for ^{60}Ni and partially for ^{62}Ni and ^{58}Fe, depending on the prior neutron excess η.

Figure 11.6 shows that the buildup of intermediate mass nuclei, like ^{38}Ar, ^{42}Ca, ^{46}Ti, and even ^{50}V, is dependent not only on the temperatures at which oxygen burns, but also on the ^{12}C(α,γ) ^{16}O reaction used in central He-burning. The faster ^{12}C(α,γ) ^{16}O rate transformed more ^{12}C into ^{16}O during He-burning. This enhanced ^{16}O abundance is almost unaffected by the subsequent C- and Ne-burning. Thus more ^{16}O fuel is available for O-burning and leads to a longer time scale. Thus reaction sequences can proceed farther and reactions with long time scales can become effective.

Table 11.6
Oxygen Burning

(a) basic reactions

$$^{16}\text{O} \left[^{16}\text{O}, \begin{matrix} p \\ \alpha \\ n \end{matrix} \right] \begin{matrix} ^{31}\text{P} \\ ^{28}\text{Si} \\ ^{31}\text{S}(\beta^+) \ ^{31}\text{P} \end{matrix}$$

$^{31}\text{P}(p,\alpha) \ ^{28}\text{Si}(\alpha,\gamma) \ ^{32}\text{S}$

$^{28}\text{Si}(\gamma,\alpha) \ ^{24}\text{Mg}(\alpha,p) \ ^{27}\text{Al}(\alpha,p) \ ^{30}\text{Si}$

$^{32}\text{S}(n,\gamma) \ ^{33}\text{S}(n,\alpha) \ ^{30}\text{Si}(\alpha,\gamma) \ ^{34}\text{S}$

$^{28}\text{Si}(n,\gamma) \ ^{29}\text{Si}(\alpha,n) \ ^{32}\text{S}(\alpha,p) \ ^{35}\text{Cl}$

$\phantom{^{28}\text{Si}(n,\gamma) \ ^{29}\text{Si}} \diagdown (p,\gamma) \ ^{30}\text{P}(\beta^+) \ ^{30}\text{Si}$

 electron captures

$^{33}\text{S}(e^-,\nu) \ ^{33}\text{P}(p,n) \ ^{33}\text{S}$

$^{35}\text{Cl}(e^-,\nu) \ ^{35}\text{S}(p,n) \ ^{35}\text{Cl}$

(b) massive stars ($M_\alpha = 16 \ M_\odot$)

$^{32}\text{S}(\alpha,\gamma) \ ^{36}\text{Ar}(\alpha,p) \ ^{39}\text{K}$

$\phantom{^{32}\text{S}(\alpha,\gamma) \ ^{36}\text{Ar}} \diagdown (n,\gamma) \ ^{37}\text{Ar}(\beta^+) \ ^{37}\text{Cl}$

$^{35}\text{Cl}(\gamma,p) \ ^{34}\text{S}(\alpha,\gamma) \ ^{38}\text{Ar}(p,\gamma) \ ^{39}\text{K}(p,\gamma) \ ^{40}\text{Ca}$

$\phantom{^{35}\text{Cl}(\gamma,p)} \diagdown (e^-,\nu) \ ^{35}\text{S}(\gamma,p) \ ^{34}\text{S} \diagdown (\alpha,\gamma) \ ^{42}\text{Ca} \begin{matrix} (\alpha,\gamma) \ ^{46}\text{Ti} \\ (\alpha,p) \ ^{45}\text{Sc}(p,\gamma) \ ^{46}\text{Ti} \end{matrix}$

(c) lower mass stars ($M_\alpha = 4 \ M_\odot$)

$^{31}\text{P}(e^-,\nu) \ ^{31}\text{S}$ $^{31}\text{P}(n,\gamma) \ ^{32}\text{P}$

$^{32}\text{S}(e^-,\nu) \ ^{32}\text{P}(p,n) \ ^{32}\text{S}$

$^{33}\text{P}(p,\alpha) \ ^{30}\text{Si}$

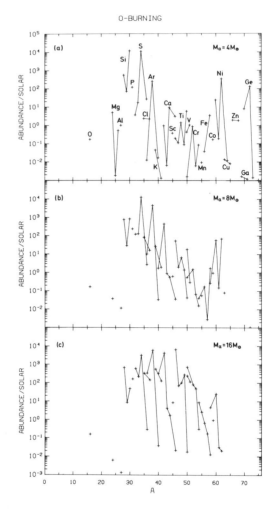

Fig. 11.5. Abundances after core O-burning. Main burning products are ^{28,30}Si and ^{34}S. With increasing burning temperatures for more massive stars, important quantities of ^{38}Ar, ^{42}Ca, and ^{46}Ti can also be created. At the same time photodisintegration reactions destroy all preexisting heavy nuclei above the Fe-peak.

11.7 Silicon Burning

During Si-burning, high temperatures allow reaction links (by neutron, proton, alpha reactions and photodisintegrations) to almost all nuclei so that a nuclear statistical equilibrium is reached for strong and electromagnetic

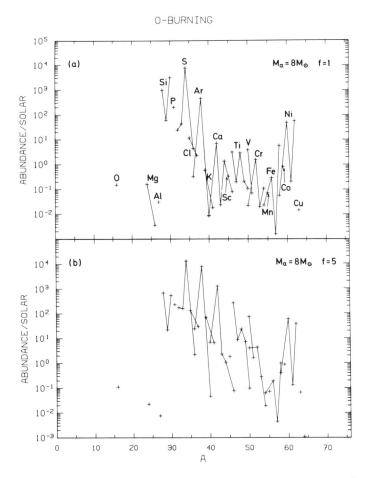

Fig. 11.6. Figure 11.5 showed that increasing temperatures, corresponding to more massive stars, lead to a larger enhancement of medium nuclei. The same can be accomplished by different amounts of ^{16}O fuel. When an enhanced $^{12}C(\alpha,\gamma)$ ^{16}O rate is used (here $f = 5$), larger amounts of ^{16}O are produced in He-burning, leaving more fuel for O-burning. This causes a more intense O-burning and results in a stronger enhancement of medium nuclei.

reactions. Before this total nuclear statistical equilibrium is gained, however, QSE clusters are formed which are connected by reactions which are not yet in an equilibrium. This effect, discussed by Woosley, Arnett, and Clayton (1973) for explosive Si-burning, also takes place in hydrostatic burning. An important dividing line between two QSE clusters is the proton magic number

$Z = 20$, which acts for charged particle induced reactions like neutron magic numbers in the r-process and prevents the strong buildup of nuclei beyond the magic number. Small reaction Q-values cause the compound nucleus to form at low excitation energies E so that γ-transmission coefficients are small ($T_\gamma \sim E^3$) and only few final levels are available ($\rho(E) \sim \exp(\sqrt{E})$). The resulting cross sections and reaction rates are small; this region of the nuclear chart acts as a bottleneck between two QSE clusters with proton numbers $12 \leq Z < 20$ and $22 \leq Z \leq 28$, respectively. The reactions which bridge this bottleneck are listed in table 11.7. Members of the first QSE group are denoted by a square ($\square$) and members of the second QSE group by a circle (O). High temperature burning (e.g., $M_\alpha = 16\ M_\odot$) bridges the gap at the proton-rich side of the stability line, while low temperature burning (e.g., $M_\alpha = 4\ M_\odot$) tends to produce neutron-rich Ca-isotopes before proton captures occur. This complex behavior shows that a simple ^{45}Sc(pc,γ) ^{46}Ti link (Weaver, Woosley, and Fuller 1983) is not generally adequate for all burning conditions. Si-burning can be divided into three substages: (a) transformation of ^{28}Si into ^{30}Si, which is dependent on the neutronization history and burning temperature (see also Woosley et al. 1984), (b) buildup of the light QSE group ($12 \leq Z < 20$), and (c) bridging the magic proton number $Z = 20$, buildup of the heavy QSE group ($20 < Z < 28$) and total nuclear statistical equilibrium.

The initial phase when ^{28}Si is transformed into ^{30}Si depends very much on the conditions. In the case of the initial 4 $M_\odot$ He-core, which had strong electron capture in oxygen burning (a final neutron excess of $\eta \approx 3.5 \times 10^{-2}$), the ^{30}Si/^{28}Si ratio reaches a maximum of 4 when the total mass is still mostly Si. For the 8 $M_\odot$ He-core, this maximum ratio is about 1, and for the 16 $M_\odot$ only 1/9. In the last case the neutron excess η before Si-burning is less than 10^{-2}; QSE with a smaller neutron/proton ratio prefers less neutron-rich nuclei. The reactions for this transformation from ^{28}Si to ^{30}Si include ones already given in O-burning, such as

^{28}Si(γ,α) ^{24}Mg(α,p) ^{27}Al(α,p) ^{30}Si,
^{28}Si(n,γ) ^{29}Si(p,γ) ^{30}P(n,p) ^{30}Si,
^{29}Si(n,γ) ^{30}Si,

and

^{29}Si(n,α) ^{26}Mg(p,γ) ^{27}Al(α,p) ^{30}Si.

Table 11.7
Reaction Links between the Two QSE-Groups in Si-Burning

(a) basic reactions

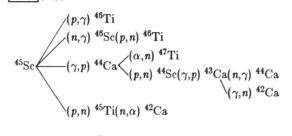

$\boxed{^{38}\text{Ar}}$ (α,γ)
$\boxed{^{39}\text{K}}$ (α,p) $^{42}\text{Ca}(\alpha,p)\ ^{45}\text{Sc}$

^{45}Sc
- $(p,\gamma)\ ^{46}\text{Ti}$
- $(n,\gamma)\ ^{46}\text{Sc}(p,n)\ ^{46}\text{Ti}$
- $(\gamma,p)\ ^{44}\text{Ca}$
  - $(\alpha,n)\ ^{47}\text{Ti}$
  - $(p,n)\ ^{44}\text{Sc}(\gamma,p)\ ^{43}\text{Ca}(n,\gamma)\ ^{44}\text{Ca}$
    - $(\gamma,n)\ ^{42}\text{Ca}$
- $(p,n)\ ^{45}\text{Ti}(n,\alpha)\ ^{42}\text{Ca}$

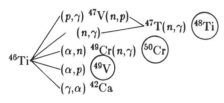

^{46}Ti
- $(p,\gamma)\ ^{47}\text{V}(n,p)$
- (n,γ) $\longrightarrow ^{47}\text{T}(n,\gamma)\ \boxed{^{48}\text{Ti}}$
- $(\alpha,n)\ ^{49}\text{Cr}(n,\gamma)\ \boxed{^{50}\text{Cr}}$
- $(\alpha,p)\ \boxed{^{49}\text{V}}$
- $(\gamma,\alpha)\ ^{42}\text{Ca}$

(b) additional reactions in high temperature burning

$^{47}\text{V}(p,\gamma)\ ^{48}\text{Cr}(n,\gamma)\ ^{49}\text{Cr}(n,\gamma)\ \boxed{^{50}\text{Cr}}$
$^{47}\text{Ti}(p,\gamma)\ ^{48}\text{V}(n,p)\ \boxed{^{48}\text{Ti}}$

(c) important reactions in low temperature burning

$\boxed{^{40}\text{Ar}}$ (α,n)
$^{42}\text{Ca}(n,\gamma)$ $\longrightarrow ^{43}\text{Ca}(n,\gamma)\ ^{44}\text{Ca}(n,\gamma)\ ^{45}\text{Ca}(n,\gamma)\ ^{46}\text{Ca}$

$^{46}\text{Ca}(p,\gamma)$
$^{46}\text{Sc}(n,\gamma)$ $\longrightarrow ^{47}\text{Sc}(p,n)\ ^{47}\text{Ti}(n,\gamma)\ \boxed{^{48}\text{Ti}}$

$^{46}\text{Ca}(n,\gamma)\ ^{47}\text{Ca}(n,\gamma)\ ^{48}\text{Ca}(p,n)\ ^{48}\text{Sc}$
- $(p,n)\ \boxed{^{48}\text{Ti}}$
- $(n,\gamma)\ ^{49}\text{Sc}(p,n)\ \boxed{^{49}\text{Ti}}$

$\boxed{}$ members of the first QSE group

$\bigcirc$ members of the second QSE group

Electron captures change the neutron excess continuously, and force the QSE to change according to the actual neutron excess.

In addition to the nuclei with high electron capture rates mentioned for O-burning (i.e. ^{33}S, ^{35}Cl, ^{31}P, and ^{32}S), ^{37}Ar and ^{36}Ar become important in the light QSE group at higher temperatures. In the heavy QSE group the nuclei which have the strongest electron capture are ^{51}Cr, ^{53}Mn, ^{54}Fe, ^{54}Mn, ^{55}Fe, ^{55}Co, ^{56}Co, ^{57}Co, and ^{59}Ni. The calculations were stopped when ^{28}Si by mass dropped below 1%. At this condition, the neutron excess for the different initial He-core masses is similar ($\eta = 6$–8×10^{-2}), as are the central temperatures ($T_9 \approx 3.5$). The densities, however, differ by factors of 1.5–2, respectively (see figure 11.2). The most abundant nuclei are ^{52}Cr and ^{56}Fe according to the different nuclear statistical equilibrium conditions. The ratio ^{52}Cr/^{56}Fe has the values ≈ 2.5 ($M_\alpha = 4~M_\odot$), ≈ 0.8 ($M_\alpha = 8~M_\odot$), and ≈ 0.6 ($M_\alpha = 16~M_\odot$), respectively. The abundances relative to solar are displayed in figure 11.7, which shows the overabundance of Fe-peak elements for $50 \leq A \leq 60$. The low mass core ($M_\alpha = 4~M_\odot$), having slightly higher neutron excess and larger density, gives a larger overabundance for the neutron-rich nuclei within one isotopic chain; this effect reverses for the 16 $M_\odot$ He-core, with $M_\alpha = 8~M_\odot$ taking an intermediate position.

In Si-burning the abundances even of the main products depend very sensitively on small changes of neutron excess, temperature, and density.

11.8 Evolution of Global Parameters

11.8.1 Energy Generation and Energy Losses

Figure 11.8 gives the behavior of all energy generation and loss terms throughout all burning stages. Since the evolution speeds significantly in late burning stages, the evolution is displayed by time steps. All quantities are central values except for the average photon luminosity L/M (the total energy loss by photons divided by the total mass), and given in ergs g^{-1} s^{-1}. They are the terms appearing in the energy equation

$$\dot{E} + p_c \dot{V}_c = 1.608 \times \left(\epsilon_c \frac{<\epsilon>}{\epsilon_c} - \frac{L}{M} - S_{\nu,c} \frac{<S_\nu>}{S_{\nu,c}} - S_{\nu\beta,c} \frac{<S_{\nu\beta}>}{S_{\nu\beta,c}} \right)$$

(eqs. [3] and [4]) which govern the evolution of the stellar core. Because each of the quantities has a different temperature and density dependence (and therefore radial dependence in the core), the curves for the average values

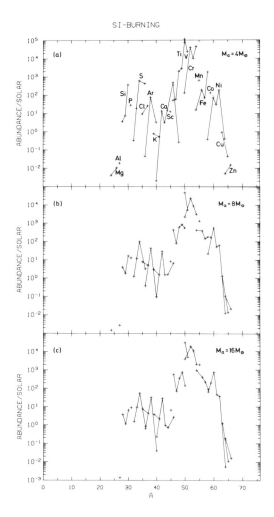

Fig. 11.7. Abundances in Si-burning as Si drops below 1% by mass. At the end of Si-burning nuclear statistical equilibrium (NSE) is reached, and thus the abundances only reflect the burning conditions (T and ρ) and the neutron excess η. The lower mass stars experience more electron captures on nuclei in O- and Si-burning and have a larger neutron excess η. The influence can be seen when regarding cases (a), (b), and (c). There is a gradual change from favoring neutron-rich isotopes of one element (in case of large η's) to favoring the more neutron deficient isotopes (for smaller η's).

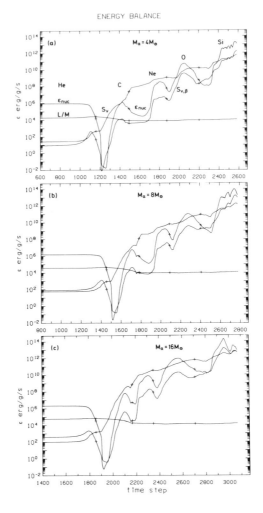

Fig. 11.8. Energy generation and loss terms for all evolutionary phases. Except for L/M (averaged photon luminosity) all quantities are given as central values. Because of an enormous speedup in late burning stages, the evolution is displayed as a function of time step number rather than time. The individual burning stages can be recognized as local maxima in the nuclear energy generation ϵ_{nuc}. From C-burning on, neutrino losses (due to pair annihilation, photo, and plasmon neutrinos) dominate over the photon radiation losses L/M. Neutrino losses due to β-decays and electron captures on nuclei $S_{\nu\beta}$ become comparable to S_ν for all stellar masses in Si-burning. For lower stellar masses ($M_\alpha = 4\ M_\odot$) this is already the case in the O-burning stage.

would look somewhat different when including the factors $<\epsilon>/\epsilon_c$ of equation (4) but reveal a similar general behavior.

The subsequent burning stages are easily recognized as local maxima in the nuclear energy generation ϵ_c. Through He-burning the main energy loss term is the photon luminosity (see eq. [8]). After He-burning, neutrino losses by pair annihilation neutrinos, photo neutrinos, and plasmon neutrinos dominate the energy losses and the time scale of core evolution (Beaudet, Petrosian, and Salpeter 1967; Ramadurai 1976). An additional neutrino loss term is introduced by neutrinos accompanying β-decay or electron capture (Fuller, Fowler, and Newman 1982). Unlike the other neutrino losses, this term follows the nuclear energy generation. In some cases there is a slight delay when the unstable burning products decay on a long time scale (see, e.g., at the end of He-burning when the decay of long-lived s-nuclei sets in).

For $M_\alpha = 4\ M_\odot$ O-burning, neutrino losses due to electron capture are comparable to the BPS neutrino losses, as the low temperature and high density favor electron captures. In Si-burning neutrino losses due to electron capture become the dominant energy loss term for all three initial He-core masses. The wiggling of ϵ_{nucl} and $S_{\nu\beta}$ in Si-burning indicate the silicon flashes. It should, however, be noticed that this simple model with the assumptions of complete mixing and hydrostatic equilibrium will give a poor description of such effects.

11.8.2 Neutron Excess η

Although a few reactions might accurately describe the main energy generation in a specific burning stage, η reflects the subtle details and small abundances of certain nuclei on the scale of 10^{-2} in mass. Nonetheless it is important to the nucleosynthesis yields and the late burning stages in which small changes in the electron abundance Y_e have large effects on the dynamic behavior of the final core collapse.

The evolution for different He-cores can be seen in figure 11.9. The main burning products of H-burning are ^{4}He and ^{14}N, both nuclei with equal proton and neutron numbers. Therefore η starts at a very low level, close to 10^{-4}. In He-burning the main increase is due to the β^+-decay in the reaction sequence ^{14}N(α,γ) ^{18}F(β^+) ^{18}O(α,γ) ^{22}Ne$[(\alpha,n)$ ^{25}Mg] of a nucleus with an abundance of the order 10^{-3}. This is almost identical for all three initial He-core masses, but the neutron production via ^{22}Ne(α,n) ^{25}Mg is temperature dependent and

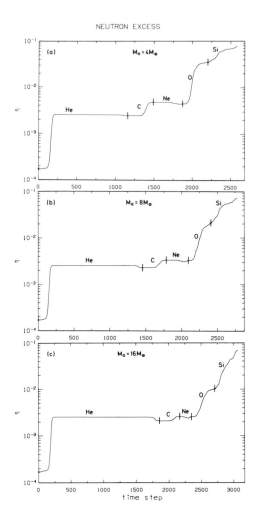

Fig. 11.9. The evolution of the neutron excess η during all evolutionary phases. $\eta = \Sigma_i (N_i - Z_i) Y_i$ is only changed by β-decays or electron captures on nuclei. The increase during He-burning is due to the reaction sequence $^{14}N(\alpha,\gamma) \, ^{18}F(\beta^+) \, ^{18}O(\alpha,\gamma) \, ^{22}Ne$ in all given cases. Further details are discussed in the text. The most pronounced differences are: (i) for low temperature C-burning the reaction sequence $^{12}C(p,\gamma) \, ^{13}N(\beta^+) \, ^{13}C(\alpha,n) \, ^{16}O$ leads to a strong increase of η, (ii) under low temperature and high density conditions, electron capture on nuclei occur even in O-burning and cause a strong increase of η.

most effective for more massive stars. The β^--decays during s-processing decrease η, and consequently this effect is bigger for $M_\alpha = 16\ M_\odot$.

In C-burning an η-enhancement generally occurs only via the low flux reaction sequence

$$^{20}\text{Ne}(n,\gamma)\ ^{21}\text{Ne}(p,\gamma)\ ^{22}\text{Na}(\beta^+)\ ^{22}\text{Ne}(\alpha,n)\ ^{25}\text{Mg}(p,\gamma)\ ^{26}\text{Al}(\beta^+)\ ^{26}\text{Mg}$$

(see table 11.4) which leads to a small enhancement in the case of $M_\alpha = 16\ M_\odot$. For low temperature, high density C-burning, the reaction sequence $^{12}\text{C}(p,\gamma)\ ^{13}\text{N}(\beta^+)\ ^{13}\text{C}(\alpha,n)\ ^{16}\text{O}$ (which also supports s-processing) is effective and an increase of η by more than a factor of 2 is achieved for $M_\alpha = 4\ M_\odot$. The neutron excess is little affected by Ne-burning. In O-burning one of the outgoing channels of the $^{16}\text{O} + ^{16}\text{O}$ fusion reaction, $^{16}\text{O}(^{16}\text{O},n)\ ^{31}\text{S}$, leaves an unstable nucleus which decays by β^+-decay and again leads to an increase of η. Electron capture on nuclei is also effective during O-burning of lower mass stars ($M_\alpha = 4\ M_\odot$), and the strongest increase of η (by a factor of 8) is for $M_\alpha = 4\ M_\odot$.

During Si-burning, electron capture is important for all He-core masses and most effective in heavier cores (see figure 11.8). Therefore the three cases almost converge to a value between 6 and 8×10^{-2} after ^{28}Si is depleted to less than 1%.

11.9 Preliminary Application to Hydrostatic Shell Burning and Pre-Supernova Abundances

Extended convective cores develop in H- and He-burning. Compared to the time scales of subsequent core burning phases, the outward movement of the hydrogen and helium burning shells is slow. In the pre-supernova stage the matter inside those shells can be regarded as consisting essentially of core (hydrogen and helium) burning products. There are interesting small effects due to shell burning which we ignore at present for simplicity. Starting with carbon burning, cores of the size of a Chandrasekhar mass (M_{Ch}) are formed (which is also thought to be the size of the neutron star remnant after a type II supernova explosion). Thus the ejecta after a SN explosion contain only *shell* burning products of C-, Ne-, O-, and Si-burning.

A simple way to predict the abundance yields is to take the abundance of material in the H- and He-burning shell from typical core burning conditions, while matter in the convective C-, Ne-, O-, and Si-burning shells is considered

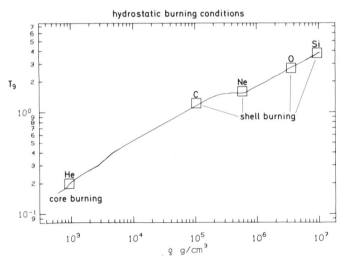

Fig. 11.10. To predict the abundances of a 20–25 $M_\odot$ star before core collapse after Si-burning, a simplified onion shell model of the star was assumed. The (later ejected) matter outside the remaining neutron star consists of *core* H- and He-burning products and *shell* C-, O-, Ne-, and Si-burning products; those burning stages took place successively. The conditions are similar to those of an 8 $M_\odot$ He-core from Arnett (1972a) and (1977).

to be from typical shell burning conditions. Those conditions (ρ, T) and the positions of the convective burning shells in the mass coordinate can be taken from evolutionary calculations (e.g., Arnett 1977; Weaver, Zimmerman, and Woosley 1978). Figure 11.10 shows the typical burning conditions for which the nucleosynthesis calculations of this paper were performed; they are close to the conditions in the evolutionary calculations of Arnett (1977). The method developed for core burning conditions was (mis)used here by choosing appropriate "effective" core masses to simulate the displayed shell burning conditions. The abundances after core hydrogen burning were taken from Arnould and Nørgaard (1978).

These nucleosynthesis calculations took into account the conditions for the evolution of an 8 $M_\odot$ He-core regarding the position in mass coordinates, temperatures, and densities as described in detail by Arnett (1972–1977). Following this recipe and adding up the results of several burning stages leads to an overabundance curve of elements as given in figure 11.11 when the original $^{12}C(\alpha,\gamma)\,^{16}O$ rate of Fowler, Caughlan, and Zimmerman (1975) is applied. The original stellar evolutionary calculations by Arnett used a value roughly

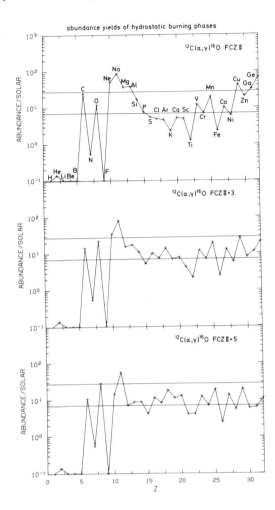

Fig. 11.11. Overproduction factors with respect to solar for the elements up to Ge, produced during hydrostatic burning phases for the (later ejected) matter outside the remaining neutron star, before the onset of core collapse in a 20–25 $M_\odot$ star (8 $M_\odot$ He-core). (a), (b), and (c) result from multiplying the $^{12}C(\alpha,\gamma)\,^{16}O$ rate by Fowler et al. (1975) by a factor of 1, 3, and 5. The curve flattens with an enhanced $^{12}C(\alpha,\gamma)\,^{16}O$ rate resulting in *relative* abundances close to solar. The additional consideration of shell He-burning products (this approximation takes only *core* He-burning products into account, see figure 11.10) would mainly affect the abundances of ^{4}He, ^{12}C, ^{16}O, ^{19}F, ... and *s*-process nuclei.

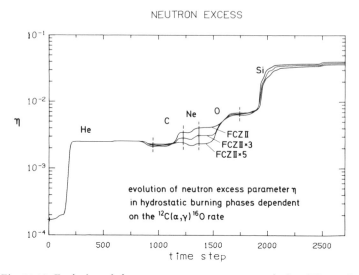

Fig. 11.12. Evolution of the neutron excess parameter η during different hydrostatic core and shell burning stages (see also figure 11.10). An enhanced $^{12}C(\alpha,\gamma)\,^{16}O$ rate results in a lower value of η in oxygen burning, which better meets the requirements for producing a solar abundance pattern in explosive oxygen burning.

twice as large. The two solid lines include an area with overproductions from 7 to 28 (14 with an uncertainty of a factor of 2) which was indicated as an average overproduction for a 25 $M_\odot$ star ($\sim$9 $M_\odot$ He-core) by Woosley and Weaver (1982), who used the Fowler rate. Both calculations give an s-shaped curve in Z or A. The carbon burning products (including Ne, Na, and Mg) are enhanced too much while O and Si shell burning products (including P, S, Cl, Ar, K, Ca, Sc, and Ti) are too low. The reason that V, Cr, and Mn behave differently from what Woosley and Weaver found is probably due to our different choice of the mass cut (1.32 $M_\odot$) for the neutron star remnant. This reflects a fundamental uncertainty. How might we obtain a constant overabundance which produces relative elemental abundances in solar ratios? An enhanced $^{12}C(\alpha,\gamma)\,^{16}O$ reaction rate would enhance the $^{16}O/^{12}C$ ratio after core He-burning. This leaves less fuel for carbon burning and more fuel for shell oxygen burning and eventually silicon burning. Another possibility would be that more massive stars produce mainly explosive oxygen burning products (see Woosley, Axelrod, and Weaver 1984).

However, the recent experimental result by Kettner et al. (1982) suggests an enhancement by a factor of 3–5 for $^{12}C(\alpha,\gamma)$ ^{16}O over Fowler et al. (1975), which supports the first possibility. A reevaluation of the Münster data by Langanke and Koonin (1984) results in roughly a factor of 3; the "generally believed" present uncertainty is within a factor of 2–4. Figures 11.11b and 11.11c show results when the $^{12}C(\alpha,\gamma)$ ^{16}O rate was enhanced by a factor of 3 and 5, respectively. The overabundance curve flattens out remarkably, except for ^{23}Na. This could reflect uncertainties in the reactions $^{12}C(^{12}C,p)$ ^{23}Na, $^{12}C(^{12}C,\alpha)$ ^{20}Ne, and/or $^{23}Na(p,\alpha)$ ^{20}Ne or indicate that a slightly higher temperature is needed for shell carbon burning. The nucleosynthesis calculations were all performed with the same stellar structure, an approximation which must be checked by evolutionary calculations. With various $^{12}C(\alpha,\gamma)$ ^{16}O rates, figure 11.12 shows the neutron excess parameter η and its change during subsequent evolutionary phases. As discussed for the core evolution (see § 11.8) there is a large increase in η during He-burning due to $^{18}F(\beta^+)$ ^{18}O, followed by a minor decrease due to β^--decay of s-process products. The increase in later burning stages is given by β^+-decays of proton-rich nuclei (produced by fusion reactions of matter with essentially $Z = N$ while the line of stability bends toward more neutron-rich nuclei). Having less fuel for carbon burning leads to a smaller increase of η. Thus a fast $^{12}C(\alpha,\gamma)$ ^{16}O rate results in a small η after carbon burning which better matches the requirements by Woosley, Arnett, and Clayton (1973) for reproducing solar abundances in O-burning.

We conclude that a changed $^{12}C(\alpha,\gamma)$ ^{16}O rate results in elemental overabundance ratios for hydrostatic burning phases which are almost constant from O to Ge. A 20–25 $M_\odot$ star, regarded as an average type II supernova progenitor, could actually give a solar abundance pattern. This, however, has to be checked in two ways: (a) Does a consistent stellar evolution calculation change the stellar structure and give different burning conditions for different $^{12}C(\alpha,\gamma)$ ^{16}O rates? (b) How much does explosive processing change the integrated abundances of the hydrostatic burning products? Arnett and Wefel (1978) found negligible differences in shell and explosive C-burning; Weaver and Woosley (1980) only show discrepancies in the Ti and Fe abundances which are also needed here to improve the agreement with a solar abundance pattern. In any case, some explosive burning is needed for the isotopic abundance patterns.

11.10 Summary

We present a simple model which treats nucleosynthesis in hydrostatic core burning stages in detail. This was done with a full nuclear reaction network containing 254 nuclear species up to ^{74}Ge, and thus revealing a much larger variety of information than small networks which are designed to represent energy generation in stellar evolution codes. Core He-burning is also of interest with respect to ejecta of type II supernova events as He-burning cores have a large size and matter in the outer parts will be unaffected by later burning stages. The strong dependence of the final ^{12}C/^{16}O ratio and s-abundances on recent but still uncertain changes of the ^{12}C(α,γ) ^{16}O reaction rate (Kettner et al. 1982; Langanke and Koonin 1984) is shown by results for three different cases.

At the low temperatures ($T_9 < 0.8$) important for stars less massive than ~ 15 $M_\odot$, the early findings of Arnett and Truran (1969) are confirmed that the reaction sequence ^{12}C(p,γ) ^{13}N(β^+) ^{13}C(α,n) ^{16}O, sustained by protons and α-particles from the ^{12}C + ^{12}C fusion reaction, is a powerful neutron source. For the subsequent Ne to Si core burning the application of an essentially complete reaction network allows testing of minimum size reaction networks to be used in Ne- and O-burning. Those must follow variables like the neutron excess η accurately, which is one of the most important quantities during core evolution. These reaction networks have to be larger than the ones used in present evolutionary calculations.

The treatment of Si-burning shows that the link between the two QSE groups around Si and Fe is a very subtle function of the individual burning conditions, reflecting the stellar mass and the prior neutron excess before Si-burning. Thus an approximation using only one reaction (i.e., ^{45}Sc(p,γ) ^{46}Ti) as an effective link between the two QSE groups includes uncertainties whose effects cannot easily be foreseen. The quantitative results obtained here may lead to a simple approximation for evolutionary calculations.

For an 8 $M_\odot$ He-core (~ 20-25 $M_\odot$ star) we performed the hydrostatic nucleosynthesis for shell burning conditions which were taken from stellar evolution calculations by Arnett (1977). The integrated abundances were obtained by a simple summation over the different burning shells weighted with the appropriate shell masses. This led to similar elemental overabundance curves (over solar) for the pre-explosive composition as also found by Woosley and Weaver (1982) in the ejected material after explosive nucleosynthesis. These overabundance curves show serious deficiencies and cannot well explain the solar system abundances; they have large quantities of Ne, Na, and Mg and small amounts of P, S, Cl, K, Ca, Sc, and Ti. Enhancing the $^{12}C(\alpha,\gamma)$ ^{16}O reaction according to recent measurements and calculations, and assuming that this does not seriously alter the stellar structure, leads to a highly improved pre-explosive composition, where the C-burning products, Ne, Na, and Mg, are less pronounced while the products of high temperature O-burning are more enhanced.

Whether such an "improved" composition is preserved after explosive processing in a type II supernova event has still to be verified. The composition of the final ejecta will depend on peak temperatures and densities achieved in the shock wave, as well as the neutron excess η of the initial composition. Considering earlier explosive nucleosynthesis calculations, we find that the lower neutron excess η of oxygen burning matter which results when using an enhanced $^{12}C(\alpha,\gamma)$ ^{16}O rate is closer to the initial conditions needed to fit solar system abundances.

References

Almeida, J., and Käppeler, F. 1983, *Ap. J.*, **265**, 417.

Anders, E., and Ebihara, M. 1982, *Geochim. Cosmochim. Acta*, **46**, 2363.

Arnett, W. D. 1969, *Ap. Space Sci.*, **5**, 180.

———. 1972a, *Ap. J.*, **176**, 681.

———. 1972b, *Ap. J.*, **176**, 699.

———. 1973, *Ap. J.*, **179**, 249.

———. 1974a, *Ap. J.*, **193**, 169.

———. 1974b, *Ap. J.*, **194**, 373.

———. 1977, *Ap. J. Suppl.*, **35**, 145.

———. 1978, in *Physics and Astrophysics of Neutron Stars and Black Holes* (Bologna, Italy: Soc. Italiana di Fisica).

———. 1983, *Ap. J. (Letters)*, **263**, L55.

Arnett, W. D., and Thielemann, F.-K. 1984, *Ap. J.*, submitted.

Arnett, W. D., and Truran, J. W. 1969, *Ap. J.*, **157**, 339.

Arnett, W. D., and Wefel, J. P. 1978, *Ap. J. (Letters)*, **224**, L139.

Arnould, M., and Nørgaard, H. 1978, *Astr. Ap.*, **64**, 195.

Arnould, M.; Nørgaard, H.; Thielemann, F.-K.; and Hillebrandt, W. 1980, *Ap. J.*, **237**, 931.

Beaudet, G., Petrosian, V., and Salpeter, E. E. 1967, *Ap. J.*, **150**, 979.

Cameron, A. G. W. 1973, *Space Sci. Rev.*, **15**, 121.

——. 1982, in *Essays in Nuclear Astrophysics*, ed. C. A. Barnes, D. D. Clayton, and D. N. Schramm, p. 23 (Cambridge: Cambridge University Press).

Chandrasekhar, S. 1939, in *An Introduction to Stellar Structure* (Chicago: University of Chicago Press).

Couch, R. G., Schmiedekamp, A. B., and Arnett, W. D. 1974, *Ap. J.*, **190**, 95.

Eddington, A. S. 1930, in *The Internal Constitution of the Stars* (Cambridge: Cambridge University Press).

Endal, A. S. 1975, *Ap. J.*, **197**, 405.

Fowler, W. A. 1981, private communication.

Fowler, W. A., Caughlan, G. R., and Zimmerman, B. A. 1975, *Ann Rev. Astr. Ap.*, **13**, 69.

Fowler, W. A., and Hoyle, F. 1964, *Ap. J. Suppl.*, **9**, 201.

Fuller, G. M., Fowler, W. A., and Newman, M. J. 1980, *Ap. J. Suppl.*, **42**, 447.

——. 1982, *Ap. J. Suppl.*, **48**, 279.

Gear, J. W. 1971, in *Numerical Initial Value Problems in Ordinary Differential Equations* (Englewood Cliffs, N.J.: Prentice-Hall).

Göbel, R., Ott, U., and Begemann, F. 1978, *J. Geophys. Rev.*, **83**, 855.

Hillebrandt, W. A. 1982, *Astr. Ap.*, **110**, L3.

——. 1984, in *Proceedings of the XIth Texas Symposium on Relativistic Astrophysics, Ann. N. Y. Acad. Sci.*, in press.

Howard, W. M.; Arnett, W. D.; Clayton, D. D.; and Woosley, S. E. 1972, *Ap. J.*, **175**, 201.

Hulke, G., Rolfs, C., and Trautwetter, H. P. 1980, *Z. Phys. A*, **297**, 161.

Käppeler, F.; Beer, H.; Wisshak, K.; Clayton, D. D.; Macklin, R. L.; and Ward, R. A. 1982, *Ap. J.*, **257**, 821.

Kettner, K. U.; Becker, H. W.; Buchmann, L.; Görres, J.; Kräwinkel, H.; Rolfs, C.; Schmalbrock, P.; Trautwetter, H. P.; and Vlieks, A. 1982, *Z. Phys. A*, **308**, 73.

Lamb, S. A.; Howard, W. M.; Truran, J. W.; and Iben, I. 1977, *Ap. J.*, **217**, 213.

Langanke, K., and Koonin, S. E. 1983, *Nucl. Phys. A*, **410**, 334.

Mathews, G. J., and Käppeler, F. 1984, *Ap. J.*, submitted.

Morgan, J. A. 1980, *Ap. J.*, **238**, 674.

Ramadurai, S. 1976, *M.N.R.A.S.*, **176**, 9.

Seelmann-Eggebert, W.; Pfennig, G.; Münzel, H.; and Klewe-Nebenius, H. 1981, *Chart of the Nuclides* (Kernforschungszentrum Karlsruhe).

Thielemann, F.-K. 1980, Ph.D. thesis, MPI Munich and TH Darmstadt.

Thielemann, F.-K., Arnould, M., and Hillebrandt, W. 1979, *Astr. Ap.*, **74**, 175.

Truran, J. W., and Arnett, W. D. 1970, *Ap. J.*, **160**, 181.

Truran, J. W., and Iben, I. 1977, *Ap. J.*, **216**, 797.

Ulrich, R. K. 1973, in *Explosive Nucleosynthesis*, ed. D. N. Schramm and W. D. Arnett, p. 139 (Austin: University of Texas Press).

Wagoner, R. V. 1969, *Ap. J. Suppl.*, **18**, 247.

Wagoner, R. V., Fowler, W. A., and Hoyle, F. 1967, *Ap. J.*, **148**, 3.

Wallace, R. K., and Woosley, S. E. 1981, *Ap. J. Suppl.*, **45**, 389.

Wapstra, A. H., and Bos, K. 1977, *Atomic Data Nucl. Data Tables*, **19**, 215.

Weaver, T. A., and Woosley, S. E. 1980, *Ann. N. Y. Acad. Sci.*, **336**, 335.

Weaver, T.A., Woosley, S. E., and Fuller, G. M. 1983, in *Numerical Astrophysics: In Honor of J. R. Wilson*, ed. R. Bowers et al. (Portola, Calif.: Science Books International), in press.

Weaver, T. A., Zimmerman, G. B., and Woosley, S. E. 1978, *Ap. J.*, **225**, 1021.

Woosley, S. E., Arnett, W. D., and Clayton, D. D. 1973, *Ap. J. Suppl.*, **26**, 231.

Woosley, S. E., Axelrod, T. S., and Weaver, T. A. 1984, in *Stellar Nucleosynthesis*, ed. C. Chiosi and A. Renzini (Dordrecht: Reidel), in press.

Woosley, S. E.; Fowler, W. A.; Holmes, J. A.; and Zimmerman, B. A. 1975, Caltech preprint OAP-422.

———. 1978, *Atomic Data Nucl. Data Tables*, **22**, 371.

Woosley, S. E., and Weaver, T. A. 1982, in *Essays on Nuclear Astrophysics*, ed. C. A. Barnes, D. D. Clayton, and D. N. Schramm, p. 377 (Cambridge: Cambridge University Press).

12

Neutron Source Requirements for the Helium r-Process

A. G. W. Cameron, J. J. Cowan, and J. W. Truran

Abstract

A review is presented of the restrictions we have identified on the operation of the r-process under thermal runaway or explosive conditions in helium zones in stars and supernovae. Working in the context of the steady flow approximation, we first demonstrate that the classical waiting point approximation is not valid under these conditions. We identify the $^{13}C(\alpha,n)^{16}O$ reaction as the only viable neutron source available in these environments. We then discuss the magnitude of the differences between the beta decay rates calculated from the gross theory of beta decay and those determined from the Klapdor theory. Many of our previous calculations were performed with the assumption that the appropriate beta decay rates were 5 to 10 times the rates predicted by the gross theory. In fact, the Klapdor rates are only about 2 to 5 times as fast as the gross rates. Using rates calculated directly from the Klapdor theory, we now find that considerably larger neutron source strengths are required to ensure reasonable fits to the solar system r-process abundance patterns. These larger neutron source requirements for fitting the observed abundances appear inconsistent with the postulated astrophysical environments. We suggest the need for two distinct r-process contributions, one of which involves the local rearrangement of s-process nuclei into r-process nuclei and produces the abundance features at mass numbers near 104 and 160. The second process, the site of which is yet unidentified, is that which is responsible for the production of the major r-process abundance features.

12.1 Test of the Waiting Point Approximation

We have recently discussed (Cameron et al. 1983) the usefulness of a steady flow approximation to the r-process and have identified the limiting conditions for which it is applicable. In this approximation nuclei are fed into an r-process network at the bottom at a steady rate, and are subject to neutron capture, photoneutron emission, and beta decay followed, where appropriate, by one or more delayed neutron emissions. The abundances of the isotopes for each value of Z are adjusted until the number of nuclei entering that value of Z by beta decay from below becomes equal to the number leaving by beta decay to the Z above. This procedure treats neutron captures and photodisintegrations exactly and does not depend on the waiting point assumption (defined below). Cameron et al. (1983) utilized the steady flow approximation together with some improved neutron capture cross sections and with Klapdor's new beta decay rates to draw conclusions about the environment responsible for producing the observed r-process abundance distribution in solar system material. It was found that the neutron number density which provides the best fit to the observed r-process abundance peaks in the steady flow approximation lies in the range 10^{20} to 10^{21} cm^{-3} and that a characteristic freeze-out time of 0.01 to 0.1 s is needed to preserve this distribution from substantial alteration.

In a second publication (Cameron, Cowan, and Truran 1983) we have used the steady flow condition to examine the range of validity of the classical waiting point approximation in the r-process. In this approximation it is assumed that the rate of neutron capture on heavy nuclei is large compared to the rate of beta decay, such that the nuclei along an isotope chain come into a steady state (equilibrium) with photodisintegration. The nuclei then "wait" at that point, gaining and losing neutrons rapidly, until the beta decay takes place. The nucleus then is assumed to move rapidly to the next waiting point. In this approximation the shape of the r-process yield curve depends primarily on the beta decay rates at the waiting points, whereas if this approximation is not valid, the yield curve depends on an average of the neutron capture cross sections over the rather broad neutron capture–beta decay path.

In figure 12.1 we show a result for conditions typical of the helium r-process. This figure shows the ratio of the rates of neutron photodisintegrations to those of neutron captures in a steady flow network for a neutron

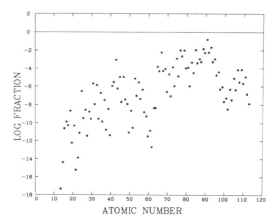

Fig. 12.1. Test of waiting point approximation for the helium r-process at a neutron number density of 10^{20} cm^{-3} and a temperature of 5×10^8 K. The ratio plotted is the ratio of the neutron photodisintegrations for a given Z relative to the direct neutron captures for the same Z.

number density of 10^{20} cm^{-3} and a temperature of 5×10^8 K. It may be seen that the photodisintegration rates are typically many orders of magnitude less than the neutron capture rates, showing that for these conditions nuclei overwhelmingly prefer to beta decay before reaching waiting points. This result is typical of all helium burning conditions up to 10^9 K; however, we find that the waiting point approximation becomes valid at temperatures of 2×10^9 K and above where explosive carbon burning occurs.

It follows that all calculations of the helium r-process, indeed, of neutron capture nucleosynthesis in general at temperatures below 10^9 K, should be carried out by following explicitly all of the neutron captures and photodisintegrations. All of the r-process calculations carried out by the three of us have done this. Those calculations which have used the waiting point approximation for the helium r-process cannot be directly compared with ours because the physical assumptions are different. It should also be noted that in the remainder of this paper we shall be stressing the importance of the competition for neutrons between heavy nuclei undergoing the r-process and lighter nuclei, many of which are produced as a consequence of the neutron production process itself. In general, those calculations which use a waiting point approximation cannot evaluate this competition.

12.2 The Neutron Source

Since it is critical to the success of the neutron capture production of heavy nuclei in the r-process that competition with lighter nuclei for the neutrons be minimized, we have found in all of our investigations that only one neutron source is plausible in the helium r-process. This is the $^{13}C(\alpha,n)^{16}O$ reaction. The neutron capture cross section of ^{16}O is extremely small, which is why heavy water is such an excellent neutron moderator. Thus, the operation of this neutron source does not increase the effectiveness of the competition of the lighter nuclei for the neutrons that are produced. On the other hand, a neutron source like $^{22}Ne(\alpha,n)^{25}Mg$ produces an end-product, ^{25}Mg, with a relatively large capture cross section. Any time that we have tried to use it as a neutron source for the r-process, we have found that the heavy nuclei only manage to capture a small fraction of the available neutrons. This results in an abundance pattern quite unlike the solar system r-process abundance distribution.

The suitability of any astrophysical environment for the operation of the helium r-process thus depends upon the mechanism available for generation of the ^{13}C neutron source. Generally this requires the production of rather a lot of ^{12}C by the triple-alpha process, followed by mixing in of some hydrogen. A great deal of latitude is not permitted in this process. The optimum mixing ratio is about 0.1 proton per ^{12}C nucleus. Most of these protons are used in making ^{13}C. If significantly more protons than this are brought in, large quantities of the ^{14}N neutron poison will build up and quickly soak up neutrons due to the large cross section for the $^{14}N(n,p)^{14}C$ reaction.

We have previously considered two possible astrophysical environments in which these events might take place. These are the helium zones in supernovae (Cowan, Cameron, and Truran 1983a) and helium core flashes in low mass stars (Cowan, Cameron, and Truran 1982, 1983b). We shall present shortly our latest, and in part preliminary, results for these scenarios.

12.3 Beta Decay Rates

We have determined that the plausibility of these helium r-process scenarios depends sensitively upon the beta decay rates for the neutron-rich heavy nuclei. Our original calculations for both cases were built upon the assumption that the correct beta decay rates were 5 to 10 times as fast as those calculated from the gross theory of beta decay (Takahashi 1971), based

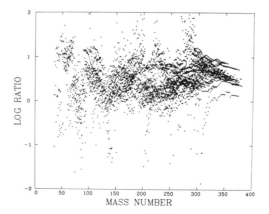

Fig. 12.2. The ratio of the Klapdor beta decay rates to the gross beta decay rates for nuclei in the r-process network.

upon early calculations of Klapdor and Oda (1980). When we later received a full set of the Klapdor rates, it turned out that they did not exceed the gross rates by quite such large factors in general. Figure 12.2 shows the ratio of the Klapdor rates to the gross theory rates; it may be seen that the former rates exceed the latter ones by factors which more typically lie in the range 2 to 5.

If one were to adopt lower beta decay rates in an r-process calculation and to change no other features of the calculation, rather dramatic effects would be evident. In the available time for the process (usually set by astrophysical considerations), the neutron capture buildup would not proceed nearly as far and in general a successful fit to the r-process abundance distribution would be transformed into a nonfit. In order to preserve the successful fit to the observed distribution, the neutron source strength would have to be increased so that the neutron capture path would be pushed more to the neutron-rich side of the valley of beta stability, where the beta decay rates would be faster, and the required r-process buildup could take place in the allowed time. The adjustment is somewhat nonlinear, since the neutron capture cross sections will be slightly reduced along the new capture path, and therefore the neutron source strength must be increased still more in order to compensate for this.

The general trend is that our previous estimates of neutron source strengths for the supernova and thermal runaway scenarios have proven to be underestimates by a factor of a few.

12.4 The Supernova Helium Zone r-Process

Successful operation of the supernova-based helium r-process requires that the supernova shock must pass outward through the envelope, cross the base of the helium shell region, produce a postshock compression there to a density of about 10^4 g cm^{-3}, and generate enough neutrons to rearrange the abundances of the heavy elements found there. Additionally, we find that a prior enhancement of the heavy elements by the s-process is required, which produces a general slope to the heavy element abundances comparable to that in the observed r-process abundances. Since in normal solar system matter the s- and r-process abundances are comparable below the 82 closed neutron shell, whereas above the shell the r-process nuclei are normally an order of magnitude more abundant than the s-process products, it follows that the prior s-process buildup this model requires is more substantial than is normally observed in the s-process. We customarily assume an s-process exposure of 25 neutrons captured per iron nucleus. It is a problem that such an exposure should be required to be common in the pre-supernova environment but not encountered in any of the sources which have mainly contributed to the observed s-process.

In the supernova process, there is only sufficient time to allow the buildup to advance an s-process abundance peak at one closed neutron shell to the next closed neutron shell in the r-process. There must be enough neutrons available to do this, for if there are not, then the initial s-process peak is left stranded in a region of mass number where no peak is seen. We are currently exploring the supernova conditions using the Klapdor beta decay rates. Figure 12.3 shows a model that provides a reasonably good fit to the observed solar system r-process distribution shown in the upper curve (from Cameron 1982). It assumes a postshock compression to 10^4 g cm^{-3} and maximum temperature of 4×10^8 K, with a ^{13}C abundance of 5.2% by mass.

While we are still in the process of exploring various physical conditions typical of the supernova environment, this disturbingly high value for the ^{13}C abundance is a typical result and is in fact a lower limit. It requires that a great deal of the helium has had to be converted to ^{12}C in the pre-supernova envelope, that hydrogen has had to be mixed into the region to produce a lot of ^{13}C, and that much of that ^{13}C has had to be used to produce neutrons for the prior s-process. The 5.2% by mass of the ^{13}C used in the subsequent r-process is what has to be *left over* after all the above has taken place.

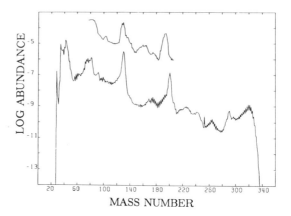

Fig. 12.3. Our best yield curve for the r-process in the supernova helium r-process. The top curve is the solar system r-process abundance curve of Cameron (1982). The assumed conditions are a postshock density of 10^4 g cm^{-3}, a maximum temperature of 4×10^8 K, and a ^{13}C initial abundance of 5.2% by mass.

This large ^{13}C abundance is necessary for the reproduction of the major features of the solar system r-process abundances. Unless further results indicate that this can be done with a much smaller neutron source, which we do not anticipate, the helium region of a supernova cannot be regarded as a promising environment for producing the main r-process abundances. It remains open to further investigation to determine whether some of the local features of the r-process can arise from this environment.

12.5 The Helium Core Flash r-Process

We have considered the r-process associated with the helium core flash to be an attractive alternative because, given enough neutrons, this process can start with just a solar system abundance distribution of nuclei and build up to observed r-process abundance distributions in a single event, thus eliminating the extensive intermediate s-process abundance distribution which is required in the supernova process but has not yet been seen in nature. We must assume that the initial composition of the core includes a significant amount of the ^{13}C neutron source. We then assume an initial temperature (usually 1.6×10^8 K) sufficient to burn the ^{13}C slowly. We let the temperature rise to reflect the heat deposited by the burning, thus producing a thermal runaway, and we artificially assume a maximum temperature achieved by the runaway which would reflect the raising of the electron degeneracy and

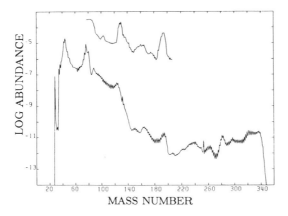

Fig. 12.4. The yield curve for the thermal runaway r-process at a density of 10^6 g cm^{-3}, a maximum temperature of 3.6×10^8 K, and a ^{13}C abundance of 3.9% by mass. The top curve is the solar system r-process abundance curve of Cameron (1982).

the expansion and subsequent cooling of the core. In our current series of runs the core density has been taken to be 10^6 K to favor higher neutron number densities and to reflect a postulated astrophysical model given below.

Figure 12.4 shows a thermal runaway result corresponding to a ^{13}C abundance of 3.9% by mass with a maximum temperature of 3.6×10^8 K. It is evident that the neutron source is quite inadequate to allow a buildup to the observed r-process abundances.

Figure 12.5 shows the results of increasing the ^{13}C abundance to 14.3% by mass, but also of raising the maximum temperature to 4.6×10^8 K. Here the agreement with observed abundances is greatly improved but the r-process buildup at the heavy end is still not quite enough. In fact, the maximum temperature is so high that the neutron source burns out too quickly.

This is corrected in figure 12.6, which also corresponds to a ^{13}C abundance of 14.3% by mass, but assumes a maximum temperature of 3.6×10^8 K. Here the buildup reproduces the general features of the r-process because it takes longer for the ^{13}C to burn, and hence the high neutron number density is maintained for a longer time.

The problem is to imagine a plausible astrophysical scenario in which the initial conditions for this solution can be produced. It is obvious that the core composition does not correspond to the initial excursion of a star into the red giant region of the H-R diagram. One must imagine a later time, when more

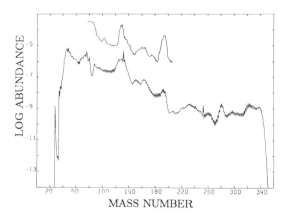

Fig. 12.5. The yield curve for the thermal runaway r-process at a density of 2×10^5 g cm^{-3}, a maximum temperature of 4.6×10^8 K, and a ^{13}C abundance of 14.3% by mass. The top curve is the solar system r-process abundance curve of Cameron (1982).

than half the helium has burned into ^{12}C and into which something like 0.2 protons have been introduced per ^{12}C nucleus (this already starts to produce uncomfortable amounts of the ^{14}N neutron poison, which we did not take into account). After this hydrogen is burned in the central region, the star must again climb the red giant branch, and since the core is probably more massive than the first time this happened, the postulated core density of 10^6 g cm^{-3} may not in itself be implausible. The thermal runaway event is then the second helium core flash episode leading to the expansion of the core. If this were to happen, only a small fraction of the r-process material produced in the core need escape to space to produce the amount of r-process material observed in nature.

We must note, however, that current studies of stellar evolution do not indicate such a behavior.

12.6 Discussion

If one rejects both these dynamic helium burning environments as possible astrophysical sites of the principal r-process, what is the outlook for finding the principal r-process? First one needs to consider whether the helium-layer neutron production reactions make any contribution at all to the r-process.

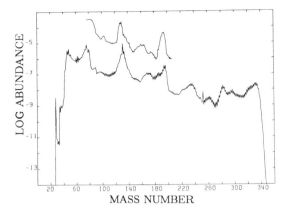

Fig. 12.6. The yield curve for the thermal runaway r-process at a density of 10^6 g cm^{-3}, a maximum temperature of 3.6×10^8 K, and a ^{13}C abundance of 14.3% by mass. The top curve is the solar system r-process abundance curve of Cameron (1982).

It seems not unlikely that they should, and possibly neutron processing associated with explosive carbon burning should as well. While we may have failed to identify the appropriate astrophysical environment, we have probably defined roughly the right nuclear physics environment for the r-process: quite dense material, a freeze-out neutron number density of 10^{20} to 10^{21} neutrons cm^{-3}, and a time scale of 0.01 to 0.1 s. The principal r-process will require these conditions, and the necessary task is to identify an environment which can produce a sufficient neutron source.

There are two significant secondary r-process abundance peaks near mass numbers 104 and 160. None of the investigations of the r-process that we have carried out in the last few years have reproduced these secondary peaks at the same time as the principal ones. It has been suggested a number of times that these secondary peaks may be due to fission from r-process nuclei formed in the transuranium region. We do not favor this suggestion for two reasons. First, in our investigations of the r-process in which we have experimented with the inclusion of fission effects, the yields of fission fragments in the range of intermediate mass numbers have always been much lower than the abundances of the nuclei there that had been built up from below: this represents a quantitative failure to reproduce the 104 and 160 peaks. Second, the 104 peak is much narrower than the 160 peak, which tends not to be characteristic of the fission process.

The question then arises whether the 104 and 160 peaks are produced by a secondary r-process. It is a characteristic of an r-process operating with a relatively weak neutron source that the abundance peaks at the positions of the closed neutron shells along the valley of beta stability formed in some earlier s-process episode may be moved upward in mass number by several units upon exposure to the neutron pulse. This is particularly true of very short pulses in which all the neutron capture takes place and the neutrons are exhausted before there is any time for beta decay. Such a condition may occur when a supernova shock enters a carbon or helium zone, leading to virtually instantaneous heating followed by expansion and cooling. It is clear that scenarios of this type need more quantitative investigation concentrating on the 104 and 160 peaks.

It is also clear that a renewed effort must be made to identify the astrophysical environment in which the principal r-process can take place. It must be an environment with an intrinsically more potent neutron source than can be generated in the helium or carbon zones that we have investigated to date. In this connection we note that Dr. Hillebrandt has informed us that, in his supernova calculations, he found the ejection of some thin mass zones in which, by compression, the total neutron to proton ratio had been raised to about a factor of 5. Also, Symbalisty (1983) found, in his investigation of the collapse of a rotating magnetized iron core, that ejection may occur of a small amount of matter in which some degree of neutronization has taken place. The investigation of the expansion of such zones would appear to be in order, since the capture of the large number of free neutrons in this environment has the potential to produce one or more features of the observed r-process distribution. In general, it would appear that the class of r-process models dealing with the expansion of highly neutronized matter from the cores of supernovae offer the most favorable conditions.

This research has been supported in part by National Science Foundation grants AST 81-19545 at Harvard University, AST 82-14964 at the University of Oklahoma, and AST 80-18198 at the University of Illinois, and by National Aeronautics and Space Administration grant NGR 22-007-272 at Harvard University.

References

Cameron, A. G. W. 1982, *Ap. Space, Sci.*, **82,** 123.

Cameron, A. G. W.; Cowan, J. J.; Klapdor, H. V.; Metzinger, J.; Oda, T.; and Truran, J. W. 1983, *Ap. Space Sci.*, **91,** 221.

Cameron, A. G. W., Cowan, J. J., and Truran, J. W. 1983, *Ap. Space Sci.*, **91,** 235.

Cowan, J. J., Cameron, A. G. W., and Truran, J. W. 1982, *Ap. J.*, **252,** 348.

——. 1983*a*, *Ap. J.*, **265,** 429.

——. 1983*b*, to appear in the Proceedings of the 3d Workshop of the Advanced School of Astronomy (Erice) on Stellar Nucleosynthesis.

Klapdor, H. V., and Oda, T. 1980, *Ap. J. (Letters)*, **242,** L49.

Symbalisty, E. M. D. 1983, Ph.D. thesis, University of Chicago.

Takahashi, K. 1971, *Prog. Theoret. Phys.*, **45,** 1466.

13

Explosive Nucleosynthesis in Carbon Deflagration Models for Type I Supernovae

Ken'ichi Nomoto

Abstract

The carbon deflagration models in accreting white dwarfs are presented as plausible models for Type I supernovae (SN I). The carbon deflagration is initiated by a relatively rapid accretion onto the C+O white dwarf and disrupts the star completely. Explosive nucleosynthesis in the deflagration wave produces 0.5–0.6 $M_\odot$ ^{56}Ni in the inner layer of the star; this amount of ^{56}Ni is sufficient to power the light curve of SN I by the radioactive decays (^{56}Ni → ^{56}Co → ^{56}Fe). In the outer layers of the star the deflagration wave synthesizes substantial amounts of intermediate mass elements, Ca, Ar, S, Si, Mg, and O; this is consistent with the spectra of SN I near maximum light. Thus the carbon deflagration model can account for many of the observed features of SN I. Moreover, the nuclear products in this model are quite complementary to nucleosynthesis in Type II supernovae.

This is in contrast to the detonation models which produce almost exclusively iron peak elements and thus cannot account for the early time spectra of SN I.

Possible SN I progenitors are discussed, namely, white dwarfs undergoing hydrogen burning near the surface and double white dwarf systems undergoing rapid accretion of helium or C+O onto the more massive white dwarf.

13.1 Introduction

13.1.1 Basic Requirements for the Type I Supernova Model

Type I supernovae (SN I) are characterized by hydrogen deficiency in their spectra near maximum light. SN I are observed in all types of galaxies including elliptical galaxies and not concentrated in spiral arms (see, e.g., Trimble 1982 for a review).

A recent breakthrough in understanding the nature of SN I has been brought about by the success of the radioactive decay model (^{56}Ni $\rightarrow$ ^{56}Co $\rightarrow$ ^{56}Fe) in reproducing the characteristic SN I light curves (Arnett 1979; Colgate et al. 1980; Chevalier 1981; Axelrod 1980b; Weaver et al. 1980; Sutherland and Wheeler 1984). This model is supported by the identification of the features in the late time spectra of SN 1972e with emission lines of Fe (Meyerott 1980; Axelrod 1980a, b). The amount of ^{56}Ni required to power the maximum light is estimated to be 0.2–1 $M_\odot$ depending on the Hubble constant (see Wheeler 1982 for a review).

The first hundred day's spectra of SN I 1981b have been compared with synthetic spectra (Branch et al. 1982, 1983); the maximum-light spectrum is well interpreted by the presence of Ca, Si, S, Mg, O, and possibly Co in the outer layer of SN I at the expansion velocity of $v_{exp} \sim 10^4$ km s^{-1}.

These results suggest the following composition structure of SN I: the inner layer, exposed at later times, contains 0.2–1 $M_\odot$ ^{56}Ni while the outer layer, observed at maximum light, is composed mainly of intermediate mass elements Ca-Si-O. This picture is consistent with the composition structure of the remnant of SN 1006 suggested from the IUE observation by Wu et al. (1983). X-ray emission from the Tycho and Kepler supernova remnants also indicates the existence of Si- and S-rich outer shells (Becker et al. 1980), although the observation of little enhancement of the iron abundance is still a problem (e.g., Shull 1982).

13.1.2 Detonation and Deflagration in Accreting White Dwarfs

The most plausible model for SN I which is consistent with many of the above features is the thermonuclear explosion of accreting white dwarfs (see Wheeler 1982 for a review and references therein). It should be noted that Hoyle and Fowler (1960) first presented the idea that SN I are triggered by the thermonuclear runaway of carbon burning in electron-degenerate cores. This idea was refined to the *carbon detonation* supernova model (Arnett 1969) and later to the *carbon deflagration* supernova model (Nomoto et al. 1976). (See Sugimoto and Nomoto 1980 for a review.)

The accreting white dwarfs in binary systems have some variations in their composition (He, C+O, O+Ne+Mg), initial mass, and age. The mass-

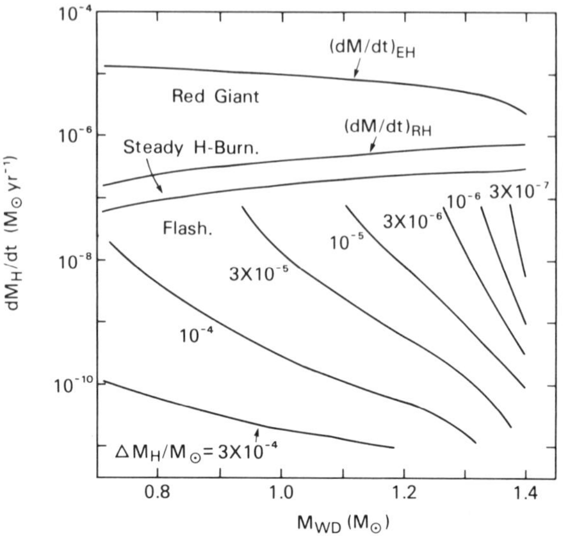

Fig. 13.1. Types and strength of hydrogen shell burning are shown as a function of the accretion rate of hydrogen-rich matter, dM_H/dt, and white dwarf mass, M_{WD}. $(dM/dt)_{EH}$ is the Eddington critical rate for hydrogen, and $(dM/dt)_{RH}$ is the growth rate of the degenerate core in red giant stars. For $(dM/dt)_{RH} \lesssim dM_H/dt \lesssim (dM/dt)_{EH}$, accretion leads to the formation of a red giant–like envelope. For $0.4\ (dM/dt)_{RH} \lesssim dM_H/dt < (dM/dt)_{RH}$, the resultant hydrogen shell burning is stable and steady. For slower accretion, the hydrogen shell burning is unstable to a flash. ΔM_H is the mass of the accreted hydrogen-rich envelope at the ignition of hydrogen.

losing companion stars fall into two classes: one is a star which transfers hydrogen-rich material to the white dwarf; the other is a white dwarf which forms through the common envelope phase of binary evolution and thus transfers helium or C+O. The latter system is called a double white dwarf system (Iben and Tutukov 1984; Webbink 1984). The combination of two component stars of the binary system therefore yields a lot of variations in the ultimate fate of mass-accreting white dwarfs.

The accreting white dwarfs can evolve to become supernovae in the following way: During the accretion, the white dwarf is compressed and heated up by the accretion and, at the same time, is cooled down by the radiation and neutrino loss. When a certain amount of mass is accreted, the density and temperature in the white dwarf interior become high enough to ignite nuclear burning depending on the composition (see, e.g., ΔM_H in figure 13.1

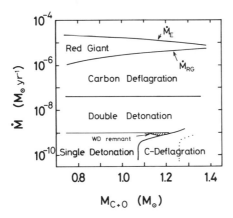

Fig. 13.2. Possible models of SN I in C+O white dwarfs, which depends on the accretion rate of helium, $\dot{M}$, and the initial mass, M_{CO}. $\dot{M}_E$ is the Eddington limit for helium. Rapid accretion ($\dot{M} \gtrsim \dot{M}_{RG}$) forms a red giant–like envelope. For $\dot{M} < 10^{-9}$ $M_\odot$ yr^{-1}, both single detonation and carbon deflagration are possible depending on M_{CO}, where the solid and dotted lines correspond to carbon ignition densities of 3.5×10^9 and 1×10^{10} g cm^{-3}, respectively. Single detonation results in either total disruption or explosion leaving a white dwarf remnant behind (*shaded region*).

for hydrogen). The condition at the ignition depends mainly on the accretion rate: for slower accretion, the interior temperature is lower and thus the ignition density is higher. This leads to a more explosive flash. Accordingly, the outcome of the flash, or types of explosion, depends on $\dot{M}$. Figure 13.1 shows such a dependence for the accretion of H-rich matter and the H-shell burning (Nomoto 1982a; see the figure legend for details; also see Sugimoto and Miyaji 1981 for a review).

Figure 13.2 shows models of supernovae induced by the accretion of *helium* onto C+O white dwarfs in the parameter plane of ($\dot{M}_{He}$, M_{CO}) where M_{CO} denotes the initial mass of the white dwarf (Nomoto 1982a, 1984b). This case corresponds to the accretion of helium from a He white dwarf as well as the growth of a helium layer through recurrence of weak hydrogen shell flashes.

For the accretion as rapid as $\dot{M} \gtrsim 4 \times 10^{-8}$ $M_\odot$ yr^{-1} in figure 13.2, both hydrogen and helium shell flashes are relatively weak and thus increase the C+O core mass (Taam 1980; Fujimoto and Sugimoto 1982). Eventually carbon is ignited at the center to become a carbon deflagration supernova.

For the slower accretion, the helium ignition is delayed to as high a density as $\rho > 2 \times 10^6$ g cm^{-3}. The resultant helium flash is strong enough to form *double detonation* waves, namely, a helium detonation wave that propagates outward and a carbon detonation wave that propagates inward (Nomoto 1980, 1982a, b; Woosley et al. 1980).

For still slower accretion, only a *single detonation* wave of helium is formed for $M_{CO} \lesssim 1.1$ $M_{\odot}$ (Nomoto 1982b; Woosley et al. 1984). If M_{CO} is larger than ~ 1.1 $M_{\odot}$, the central density of the white dwarf reaches the carbon ignition density of $\sim 4 \times 10^9$ g cm^{-3} prior to the helium ignition. Therefore a carbon deflagration supernova will result.

This detonation/deflagration type of explosion disrupts the white dwarf completely except for the single detonation that leaves a white dwarf remnant behind. The ejecta is composed mostly of ^{56}Ni, and the amount is enough to power the light curve. There is an important difference, however, between the carbon deflagration model and the detonation model. The subsonic deflagration produces substantial amounts of Ca, S, Ci, Mg, and O in the decaying phase of the deflagration (Nomoto et al. 1976; Nomoto 1980, 1981). On the other hand, the supersonic double detonations and the helium detonation in the accreting helium white dwarf produce almost exclusively ^{56}Ni (Nomoto 1980, 1982b; Woosley et al. 1980; Nomoto and Sugimoto 1977).

13.1.3 New Developments in Nucleosynthesis

These results have suggested that the carbon deflagration model is consistent with the basic composition structure of SN I ejecta, while the detonation models are not. In order to compare the models with the observations of SN I in more detail, we need to calculate detailed nucleosynthesis in the carbon deflagration model. This has recently been done by Nomoto, Thielemann, and Yokoi (1984; see Nomoto 1984b for preliminary results) using a nuclear reaction network including 205 species, where the deflagration model was calculated from the initial stage of accretion through explosion.

On the basis of such a detailed nucleosynthetic model, it is now possible to calculate synthetic spectra and compare them with the early time spectra of SN I (Branch 1984).

Another important feature of this model is that it produces significant amounts of Ca, S, Si, and Ar. These elements are apparently underproduced relative to the solar abundance in the massive star model (Woosley and

Table 13.1
Initial and Ignition Models

	Initial Models		Ignition Models		
	ρ_c(g cm^{-3})	T_c(K)	ρ_c(g cm^{-3})	$M/M_\odot$	E_{bind}(ergs)
Core	2.9×10^7	1×10^8	1.5×10^9	1.366	5.0×10^{51}
WD	3.4×10^7	1×10^7	2.6×10^9	1.378	5.3×10^{51}

Weaver 1982). Therefore it is interesting to see to what degree the abundances from SN I and SN II are complementary (Nomoto, Thielemann, and Wheeler 1984).

In this paper I will summarize the deflagration model and the associated nucleosynthesis (Nomoto, Thielemann, and Yokoi 1984) and its contribution to galactic nucleosynthesis.

13.2 Evolution of Accreting White Dwarfs

13.2.1 Accretion with Hydrogen-Helium Double Shell Burning

Nomoto, Thielemann, and Yokoi (1984) assumed that C+O white dwarfs grow through steady H and He double shell burnings, and calculated two cases of accretion. One corresponds to the C+O core in the red giant star with the accretion rate of $\dot{M} = \dot{M}_{\text{AGB}} = 8.5 \times 10^{-7}$ ($M/M_\odot$ – 0.52) $M_\odot$ yr^{-1}. The other is the white dwarf, which has been cooling down for 5.8×10^8 yr, with the accretion rate of 4×10^{-8} $M_\odot$ yr^{-1}. Compositions were assumed to be $X(^{12}\text{C}) = 0.475$, $X(^{16}\text{O}) = 0.5$, and $X(^{22}\text{Ne}) = 0.025$. Initial values of central temperature, T_c, and density, ρ_c, are summarized in table 13.1.

Figure 13.3 shows a change in the density-temperature structure of the white dwarf during accretion (*dashed lines*) and the evolutionary tracks of (ρ_c, T_c) for both the core and the white dwarf (*solid lines*). As mass is accreted, both ρ_c and T_c increase by compression. At the same time, heat conduction from the outer hot layer into the central cold layer raises T_c. The heating sources are both H-He double shell burning and the compression of the outer layer; their effects are comparable for $\dot{M}$ considered here. As a result, T_c increases to a point where neutrino loss is significant; afterward the evolution of (ρ_c, T_c) is controlled by the balance between compressional heating and neutrino loss, and thus its path depends only on $\dot{M}$.

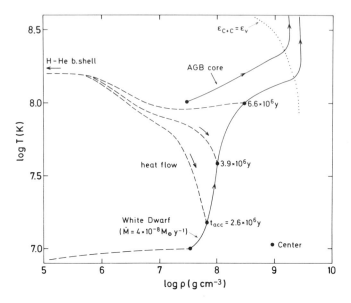

Fig. 13.3. Accretion onto the white dwarf and the core growth of the AGB star. The evolution of (ρ_c, T_c) is shown by the solid lines. Dashed lines are the structure lines of the white dwarf where heat flows from the surface into the interior. The dotted line is the ignition line of carbon burning.

Similar results have been obtained for colder white dwarfs with a crystal-lized central region. The white dwarf is first melted by the heat inflow from the surface layer and the evolutionary path of (ρ_c, T_c) merges into the same one as is determined only by $\dot{M}$ in figure 13.3. (If the initial mass of the white dwarf is more massive than, say, $\sim 1.2\ M_\odot$, the evolution of (ρ_c, T_c) in later phases will depend on the initial condition.)

When the evolutionary path of (ρ_c, T_c) reaches the dotted line in figure 13.3, carbon burning is ignited and flashes at the center. The central density at the ignition, $\rho_{c,ig}$ (summarized in table 13.1) is important for the subsequent evolution and nucleosynthesis, in particular, neutronization (§ 13.5.1). The strong screening effect due to degenerate electrons (Ichimaru and Utsumi 1983) significantly enhances the C+C reaction rate even for low temperature. Therefore the ignition density is $\rho_{c,ig} \sim 3.5 \times 10^9$ g cm^{-3} even for $\dot{M} < 10^{-9}\ M_\odot$ yr^{-1}, which is significantly lower than the previous value of $\sim 10^{10}$ g cm^{-3} for such a low temperature white dwarf (Canal and Shatzmann

Table 13.2
Accretion of C+O onto C+O White Dwarfs
and Carbon Ignition

$\dot{M}$ $(M_\odot \text{ yr}^{-1})$	$\log T_{c,0}$ (K)	$M_{\text{WD}}^{(\text{ig})*}$ $(M_\odot)$	Ignited Shell		
			$M_r(M_\odot)$	$\log \rho$	$\log T$
1×10^{-5}	7.00	1.108	1.035	6.04	8.78
4×10^{-6}	7.00	1.269	1.231	6.42	8.76
4×10^{-6}	7.67	1.281	1.242	6.45	8.76
2×10^{-6}	7.00	1.387	Center	9.58	8.03
2×10^{-6}	7.67	1.356	Center	9.10	8.40

*Ignition stage.

1976; Duncan et al. 1976; Ergma and Tutukov 1976; Nomoto 1982 a).

The above results imply that, unless the initial mass of the white dwarf is as large as $M \gtrsim 1.2\ M_\odot$, the white dwarf mass and $\rho_{c,\text{ig}}$ at the carbon ignition are fairly independent of the initial condition of the white dwarfs and $\dot{M}$. This is consistent with the uniformity of SN I.

13.2.2 Double White Dwarf System

For the double white dwarf system, their separation is small at their birth and gets smaller as angular momentum is lost from the system through gravitational radiation. When the smaller mass white dwarf fills its Roche lobe, mass transfer of helium or carbon-oxygen to the more massive white dwarf commences. Since the radius of the mass-losing white dwarf gets larger as M decreases, the system could be unstable to a dynamical time scale mass transfer. Even if it is stable, $\dot{M}$ may well exceed the Eddington limit because of a rapid decrease in angular momentum (Webbink 1984).

Quantitative studies of such rapid accretion onto the white dwarf has recently been started for the system of CO-CO white dwarf pair, i.e., accretion of C+O onto C+O white dwarfs (Nomoto and Iben 1984). The calculations have been done for three cases of accretion rate and two initial central temperatures, $T_{c,0}$, for the initial mass of $M_{\text{WD},0} = 1.0\ M_\odot$ as summarized in table 13.2. Such a rapid accretion heats up the surface layer efficiently so that the temperature rises first in the outer layers. For $\dot{M} \gtrsim 4 \times 10^{-6}$ $M_\odot$ yr^{-1}, therefore, carbon is ignited in the outer shell. (The location and density of the ignited shell, which are summarized in table 13.2, are almost

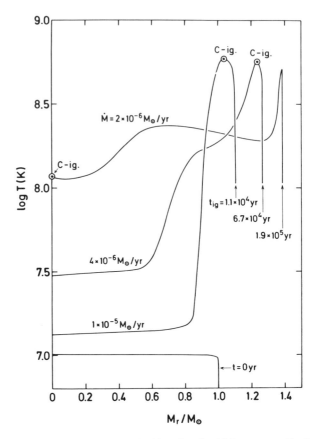

Fig. 13.4. Structure of the C+O white dwarf which accretes C+O from the white dwarf companion in the double white dwarf system. Temperature distribution at the carbon ignition is shown for various values of $\dot{M}$. Faster accretion leads to off-center carbon ignition, while central ignition occurs for slower accretion.

independent of $T_{c,0}$.) For $\dot{M} = 2 \times 10^{-6}\ M_\odot$ yr^{-1}, on the other hand, the rise in temperature in the outer layer is slower so that the carbon ignition starts at the center. Thus the critical accretion rate that discriminates between the central and off-center ignition is $\sim 3 \times 10^{-6}\ M_\odot$ yr^{-1} for $M_{\rm WD,0} = 1\ M_\odot$. The temperature distribution in the white dwarf interior at ignition is shown in figure 13.4 for three values of $\dot{M}$.

The one cycle of rise and decay of the off-center carbon flash was followed for $\dot{M} = 1 \times 10^{-5}\ M_\odot$ yr^{-1}. Almost all carbon in the convective shell is processed into O, Ne, and Mg. The maximum temperature is 1.1×10^9 K, which is too low to synthesize heavier elements. At the end of the cycle, the next carbon flash is ignited at the next inner shell because the temperature there has been increased by heat conduction. If such a burning shell continues to move inward in mass and reaches the center, the C+O white dwarf changes into an O+Ne+Mg white dwarf which will ultimately collapse (Nomoto et al. 1979). If the inward shift of the shell burning stops somewhere, on the other hand, the ultimate outcome will be a carbon deflagration at the center, but the outer layers of the white dwarf would be composed of O+Ne+Mg. The behavior of the off-center burning is therefore crucial for the double white dwarf models of SN I and deserves further study.

For the case of central flash, there is an important difference between the cases with $\dot{M} = 2 \times 10^{-6} \, M_\odot \, \mathrm{yr}^{-1}$ and $\dot{M} < \dot{M}_{\mathrm{AGB}}$. For the former case the time scale of accretion up to carbon ignition is $t_{\mathrm{ig}} = 1.9 \times 10^5$ yr, which is shorter than the heat conduction time scale of $\sim 10^6$ yr from the outer layer to the center. Therefore, the central layer is compressed almost adiabatically as seen in figure 13.5, which shows the evolutionary paths of (ρ_c, T_c). (For $T_{c,0} = 4.7 \times 10^7$ K, the central entropy decreases somewhat owing to neutrino emission.) As a result, the ignition density depends on the initial temperature of the white dwarf as shown in figure 13.5 and table 13.2. This implies that more rapid accretion does not necessarily lead to the lower ignition density, which may be important to the problem of excess neutronization as discussed in § 13.5.1.

13.3 Hydrodynamic Behavior of Carbon Deflagration Supernovae

13.3.1 Formation and Propagation of the Deflagration Wave

The central carbon flash grows into thermonuclear runaway, which incinerates the material into *nuclear statistical equilibrium* (NSE). At $\rho_c \sim 2 \times 10^9$ g cm^{-3}, however, the released nuclear energy, $\sim 3 \times 10^{17}$ ergs g^{-1}, is only 20% of the Fermi energy of degenerate electrons (see figure 2 of Nomoto 1982b) so that the thermal overpressure is too weak to form a carbon detonation wave (cf. Ivanova et al. 1974; Buchler and Mazurek 1975; Nomoto et al. 1976; Mazurek et al. 1977). The spherical geometry in the central region also damps the shock wave.

After the thermonuclear runaway in the central region, the explosive carbon-burning front propagates outward on the time scale for convective

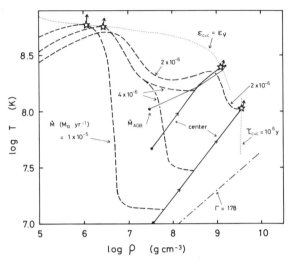

Fig. 13.5. Same as figure 13.3 but for the rapid accretion of C+O onto the C+O white dwarfs in double white dwarf systems. The dashed lines show the structure line of the white dwarf at the carbon ignition stage. Carbon is ignited at the outer shell (for $\dot{M} = 1 \times 10^{-5}$, 4×10^{-6} $M_\odot$ yr^{-1}) or at the center (for $\dot{M} = 2 \times 10^{-6}$ $M_\odot$ yr^{-1}) as indicated by the star mark. For $\dot{M} > \dot{M}_{AGB}$, the evolutionary path of (ρ_c, T_c) depends only slightly on $\dot{M}$ up to the ignition but strongly on the initial temperature. The carbon ignition at low temperature occurs when $\tau_{C+C} = c_p T / \epsilon_{C+C}$ becomes $\sim 10^6$ yr as shown by the dotted line.

energy transport across the front, because the density inversion at the front is unstable to Rayleigh-Taylor instability (e.g., Müller and Arnett 1982). A burning front which propagates at a subsonic velocity with respect to the unburnt material is called a *deflagration wave* in contrast to a detonation wave which propagates at a supersonic speed (e.g., Courant and Friedrichs 1948).

To simulate the propagation of a convective carbon deflagration wave with a 1D hydrodynamic code, Nomoto, Thielemann, and Yokoi (1984) employed the time-dependent mixing length theory of convection by Unno (1967). The mixing length, l, was taken to be $l = \alpha H_p$ where H_p is a pressure scale height and α is a parameter.

Starting from the ignition models in table 13.1, two cases with $\alpha = l/H_p$ = 0.6 (C6) and 0.8 (C8) for the core and three cases with α = 0.6 (W6), 0.7 (W7), and 0.8 (W8) for the white dwarf were calculated. Here the result of case W7 will be discussed in some detail. Figure 13.6 shows the propagation

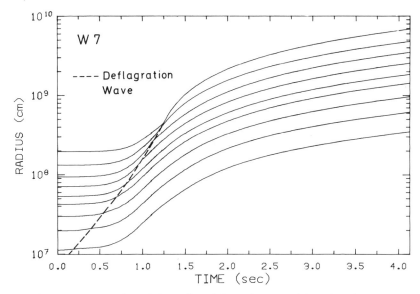

Fig. 13.6. Propagation of the carbon deflagration wave (*dashed line*) and the associated expansion of the Lagrangian shells for case W7. These shells correspond to $M_r/M_\odot = 0.007, 0.03, 0.10, 0.25, 0.41, 0.70, 1.00, 1.28$, and 1.378, from the interior to the exterior.

of the deflagration wave and the associated expansion of the Lagrangian shells. Changes in the profiles of temperature, density, and velocity are shown in figures 13.7a–13.7c.

The propagation velocity, v_{DF}, of the deflagration wave with respect to the unburnt material is slow in the early stages; e.g., $v_{DF} \sim 0.08\ v_s$ for stage 2, where v_s is the sound velocity of the burnt material. As the deflagration wave propagates outward, v_{DF} gets higher because of the increasing density jump across the front. However, at stage 7, v_{DF} is still as slow as $\sim 0.3\ v_s$. Thus it takes 1.2 s for the deflagration wave to reach the shell at $M_r = 1.3$ $M_\odot$ (figure 13.6), which is 6 times longer than the propagation of a detonation wave (Arnett 1969).

During the slow propagation of the deflagration wave, the white dwarf gradually expands to decrease the density and temperature (figures 13.7a and 13.7b). Such an expansion weakens the explosive nuclear burning and eventually quenches the carbon burning when the deflagration wave reaches $M_r = 1.3\ M_\odot$ where $\rho \sim 10^7$ g cm^{-3}.

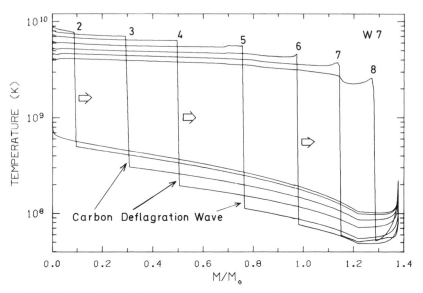

Fig. 13.7*a*. Change in the temperature profile against M_r during the propagation of the carbon deflagration wave (case W7). Stage numbers 1–9 correspond to $t(s) =$ 0.0 (#1), 0.60 (#2), 0.79 (#3), 0.91 (#4), 1.03 (#5), 1.12 (#6), 1.18 (#7), 1.24 (#8), 3.22 (#9), respectively.

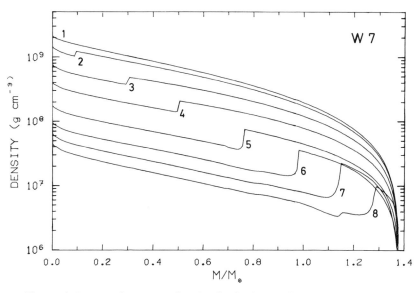

Fig. 13.7*b*. Same as figure 13.7*a* but for the density profile.

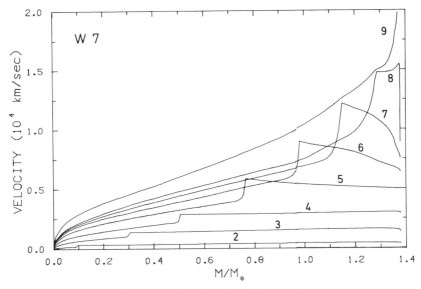

Fig. 13.7 c. Same as figure 13.7 a but for the velocity profile.

The deflagration wave compresses the material ahead of it and forms a precursor shock wave as seen from the density and velocity profiles (figures 13.7 b and 13.7 c). This precursor shock is not strong enough to ignite carbon; in other words, the deflagration wave does not grow into a detonation. When the deflagration wave arrives at the outer layer of $M_r \sim 1.3 \ M_\odot$, however, the precursor shock is strengthened appreciably owing to the steep density gradient near the surface. Whether the shock ignites carbon depends on the propagation speed of the deflagration wave and the preshock temperature. For the present models, carbon ignition in the outer layer does occur for case C8 but does not for other cases.

13.3.2 Energetics of Explosion

The energetics of the deflagration are summarized in table 13.3; the total nuclear energy release exceeds the initial binding energy of the white dwarf and the core so that the star is disrupted completely with no compact star remnant. The explosion energy is $\sim 10^{51}$ ergs, which is in good agreement with SN I. As seen from the velocity profile at stage 9, the expansion velocity of 10,000–14,000 km s^{-1} deduced from the spectra at maximum light corresponds to $M_r = 0.9$–$1.2 \ M_\odot$.

Table 13.3
Energetics of Explosion

Case	C6	C8	W6	W7	W8
$\dot{M}$	$\dot{M}_{AGB}$		$4 \times 10^{-8}\ M_\odot\ \mathrm{yr}^{-1}$		
$\alpha(\equiv l/H_p)$	0.6	0.8	0.6	0.7	0.8
E(nuclear)*	1.4	1.9	1.5	1.8	2.0
E(neutrino)*	0.018	0.020	0.033	0.035	0.037
E(kinetic)*	0.91	1.4	0.99	1.3	1.5
$M(^{56}\mathrm{Ni})/M_\odot$	0.48	0.64	0.49	0.58	0.65

*Energies in units of 10^{51} ergs.

13.3.3 Neutrino Burst

During the propagation of the deflagration wave, copious neutrinos (ν_e) are produced by electron capture processes. Total neutrino energy loss is given in table 13.3 and neutrino luminosity, L_ν, is shown in figure 13.8. If we could observe a time profile of such a neutrino burst in the future, it would provide useful information on the propagation speed of the deflagration wave.

13.4 Explosive Nucleosynthesis in the Carbon Deflagration Wave

The white dwarf material undergoes explosive burning of carbon, neon, oxygen, and silicon at the passage of the deflagration wave. Nucleosynthesis was calculated until freezing by solving the reaction network with 205 species. For the neutronization, electron capture rates by Fuller et al. (1982) were applied. Figures 13.9a and 13.9b show changes in the abundances of several species for two typical layers at $M_r = 0.73\ M_\odot$ and 1.05 $M_\odot$, respectively (Thielemann, Nomoto, and Yokoi 1984). The nuclear products of such explosive burning depend mainly on the peak temperature, T_p, and the corresponding density, ρ_p, at the deflagration wave. The final composition structure is shown in figure 13.10a for case W7 and figure 13.10b for case C6.

13.4.1 NSE Layer

For the inner layer at $M_r < 0.7\ M_\odot$, $\rho_{p,7} \gtrsim 9$ and $T_{p,9} \gtrsim 6$ so that the nuclear reactions are rapid enough to incinerate the material into almost NSE composition. Here $\rho_{p,7} \equiv \rho_p/10^7$ g cm^{-3} and $T_{p,9} \equiv T_p/10^9$ K. As the white dwarf expands, the interior temperature decreases and the composition changes into iron-peak elements. The resultant composition structure is

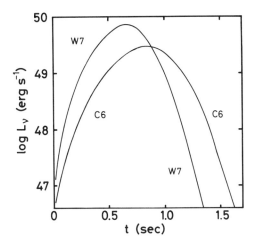

Fig. 13.8. Time profile of the neutrino luminosity, L_ν, during the propagation of a carbon deflagration wave for cases W7 and C6.

shown in figures 13.10a and 13.10b. The central layers are composed of neutron-rich iron-peak elements (^{58}Ni, ^{56}Fe, ^{54}Fe) because of electron captures at the high density; the number of electrons per baryon at the center is $Y_e = 0.46$ for case W7 and 0.47 for case C6. In the lower density region at $M_r \gtrsim 0.1\ M_\odot$, Y_e is larger so that ^{56}Ni is the dominant product.

It is noteworthy that some ^{40}Ca, ^{36}Ar, and ^{32}S remain during freezing at $0.3 \lesssim M_r/M_\odot \lesssim 0.6$. This indicates that partial breakdown of NSE occurs at relatively low temperature because nuclear reactions cannot keep pace with the change in the NSE composition due to the rapid decrease in $(\rho,\ T)$. (See the difference between the NSE [Nomoto 1984b] and the network calculation.)

13.4.2 Partially Burnt Layer

When the deflagration wave reaches the outer layers at $M_r \gtrsim 0.7\ M_\odot$, the density at the front has been significantly decreased from the initial value at $t = 0$ as a result of expansion (figure 13.7b). The peak temperature attained during explosive burning, T_p, is lower for lower density because of the larger heat capacity. Moreover, the temperature and density quickly decrease after the passage of the deflagration wave because of the accompanying rarefaction wave and the rapid expansion of the white dwarf. These effects slow down the nuclear reactions so that the material undergoes explosive burning but is

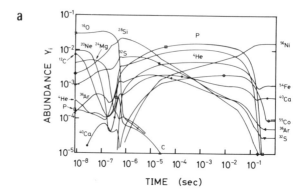

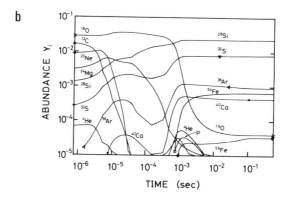

Fig. 13.9. Nucleosynthesis after the arrival of the carbon deflagration wave. Changes in the abundances, Y_i, are shown for the shells at $M_r = 0.73 \, M_\odot$ (a), and 1.05 $M_\odot$ (b), respectively, for case W7.

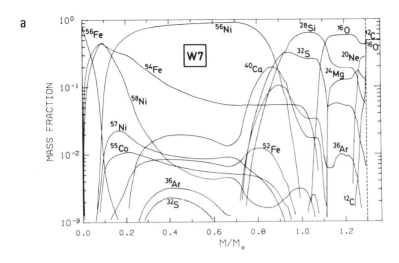

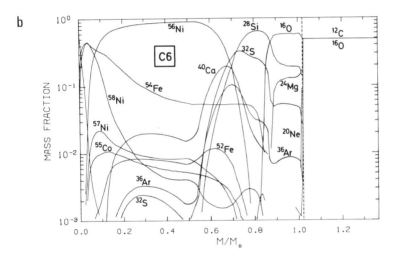

Fig. 13.10. Composition structures of the deflagrating white dwarfs for cases W7 (a) and C6 (b). At $M_r \lesssim 0.7\ M_\odot$ (a), the white dwarf undergoes incineration into NSE. In the intermediate region at $0.7 < M_r/M_\odot \lesssim 1.3$, the white dwarf undergoes partial explosive burning. In the outer layer ($M_r > 1.3\ M_\odot$), the deflagration wave is quenched.

not incinerated into NSE composition.

For the layers at $0.7 \lesssim M_r/M_\odot < 0.9$, $6 > T_{p,9} \gtrsim 5$ and $9 > \rho_{p,7} \gtrsim 4$. The reactions are not rapid enough to process the material into ^{56}Ni; in other words, this layer undergoes partial Si burning, whose products are ^{28}Si, ^{32}S, ^{36}Ar, ^{40}Ca, ^{54}Fe, ^{56}Ni, etc.

For $0.9 \lesssim M_r/M_\odot < 1.1$, $5 > T_{p,9} \gtrsim 4$ and $4 > \rho_{p,7} \gtrsim 2.5$. Explosive carbon and oxygen burning produces ^{28}Si, ^{32}S, and ^{36}Ar, but T_p is too low for Si burning to proceed. Thus this layer is rich in Si-peak elements.

For $1.1 \lesssim M_r/M_\odot < 1.25$, $4 > T_{p,9} \gtrsim 3$ so that this layer undergoes explosive carbon and neon burning, which produces ^{16}O, ^{24}Mg, ^{28}Si, etc.; oxygen burning is too slow to proceed appreciably. Finally, in the quenching phase of the deflagration, only the carbon-burning products appear for $1.25 \lesssim M_r/M_\odot < 1.3$. In the outer layer at $M_r \gtrsim 1.3\ M_\odot$, the original C+O remains unburnt.

13.5 Abundances in the Ejecta

The deflagrated white dwarfs are disrupted completely so that all the synthesized materials in figures 13.10a and 13.10b are ejected into space. The integrated mass and the ratio $<X_i/^{56}\mathrm{Fe}> \equiv (X_i/X(^{56}\mathrm{Fe}))/(X_i/(^{56}\mathrm{Fe}))_\odot$ for the stable isotopes are summarized in table 13.4 and figure 13.11 for case W7. Case C6 shows an almost identical pattern except for the smaller abundances of neutron-rich species and the larger ^{16}O abundance. Figure 13.12 shows the abundances of elements relative to solar, which is normalized to Si (case W7).

13.5.1 Iron-Peak Elements

The masses of the iron-peak elements that are ejected are 0.86 $M_\odot$ (W7) and 0.66 $M_\odot$ (C6). Among them, ^{56}Ni amounts to 0.5–0.6 $M_\odot$, which is enough to power the light curve of SN I by the decays into Co and Fe.

The abundance ratios among the iron-peak elements are generally in good agreement with solar values as has been found for the carbon detonation model (Arnett et al. 1971; Bruenn 1971). The exception is the ratio ^{54}Fe/^{56}Fe, which is enhanced by a factor of $\sim$3.9 (W7) and 3.6 (C6) relative to the solar value. This ratio depends on the distribution of the neutron excess, $\eta = 1 - 2Y_e$, in the NSE layer, which, in turn, depends on the ignition density and on the competitive processes of electron captures and the propagation of the deflagration wave; for lower $\rho_{c,\mathrm{ig}}$ and faster deflagration wave, η is smaller in

Table 13.4
Nucleosynthesis Products for Model W7

Species	Mass ($M_\odot$)	$<X_i/X_i(^{56}Fe)>*$	Species	Mass ($M_\odot$)	$<X_i/X_i(^{56}Fe)>*$
^{12}C	3.2 E-02	1.8 E-02	^{40}Ca	4.1 E-02	1.3 E 00
^{13}C	7.6 E-13	3.6 E-11	^{42}Ca	2.2 E-05	9.9 E-02
^{14}N	4.8 E-09	1.1 E-08	^{43}Ca	1.7 E-08	3.2 E-04
^{15}N	2.1 E-08	1.3 E-05	^{44}Ca	8.2 E-05	1.1 E-01
^{16}O	1.4 E-01	3.6 E-02	^{46}Ca	6.9 E-12	5.5 E-06
^{17}O	2.9 E-10	1.9 E-07	^{48}Ca	2.2 E-13	3.0 E-09
^{18}O	4.5 E-12	5.0 E-10	^{45}Sc	5.7 E-07	3.1 E-02
^{19}F	2.9 E-11	1.5 E-07	^{46}Ti	4.8 E-05	4.1 E-01
^{20}Ne	1.1 E-02	1.7 E-02	^{47}Ti	9.9 E-07	9.1 E-03
^{21}Ne	4.1 E-08	2.1 E-05	^{48}Ti	3.1 E-06	2.7 E-03
^{22}Ne	1.5 E-03	1.8 E-02	^{49}Ti	1.6 E-05	1.8 E-01
^{23}Na	1.8 E-05	9.9 E-04	^{50}Ti	1.1 E-06	1.3 E-02
^{24}Mg	2.3 E-02	8.7 E-02	^{50}V	8.4 E-09	2.1 E-02
^{25}Mg	7.8 E-06	2.2 E-04	^{51}V	2.6 E-05	1.5 E-01
^{26}Mg	2.4 E-05	6.0 E-04	^{50}Cr	3.3 E-04	9.1 E-01
^{27}Al	6.6 E-04	2.2 E-02	^{52}Cr	7.7 E-03	1.1 E 00
^{28}Si	1.6 E-01	4.5 E-01	^{53}Cr	2.8 E-03	3.3 E 00
^{29}Si	4.6 E-04	2.6 E-02	^{54}Cr	3.0 E-05	1.4 E-01
^{30}Si	4.5 E-03	3.6 E-01	^{55}Mn	8.0 E-03	1.2 E 00
^{31}P	1.7 E-04	6.4 E-02	^{54}Fe	1.4 E-01	3.9 E 00
^{32}S	8.2 E-02	4.1 E-01	^{56}Fe	6.1 E-01	1.0 E 00
^{33}S	7.8 E-04	4.7 E-01	^{57}Fe	1.1 E-02	7.5 E-01
^{34}S	1.7 E-03	1.8 E-01	^{58}Fe	2.1 E-04	9.3 E-02
^{36}S	2.3 E-08	7.2 E-04	^{59}Co	4.7 E-04	2.7 E-01
^{35}Cl	1.2 E-04	7.3 E-02	^{58}Ni	6.1 E-02	2.4 E 00
^{37}Cl	2.6 E-05	4.6 E-02	^{60}Ni	1.1 E-02	1.1 E 00
^{36}Ar	2.2 E-02	5.2 E-01	^{61}Ni	6.4 E-06	1.4 E-02
^{38}Ar	9.8 E-04	1.2 E-01	^{62}Ni	2.7 E-04	1.9 E-01
^{40}Ar	1.3 E-09	1.2 E-05	^{64}Ni	2.3 E-07	5.3 E-04
^{39}K	7.2 E-05	4.3 E-02	^{63}Cu	4.3 E-07	1.4 E-03
^{41}K	6.9 E-06	5.3 E-02	^{65}Cu	1.1 E-08	7.3 E-05

$*<X_i/^{56}Fe> \equiv (X_i/X(^{56}Fe))/(X_i/X(^{56}Fe))_\odot$.

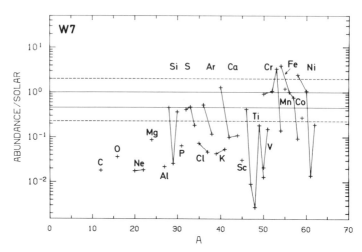

Fig. 13.11. Nucleosynthesis in the carbon deflagration model W7. The abundances of stable isotopes relative to the solar values are shown. The ratio is normalized to ^{56}Fe.

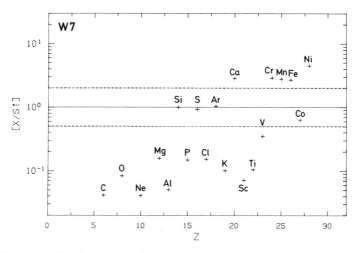

Fig. 13.12. The abundances of elements relative to the solar values (W7). The ratio is normalized to Si.

the central region. Therefore, the distribution of η or Y_e is sensitive to the treatment of deflagration wave. In view of the uncertainties involved in the theory of convection, the overabundance of ^{54}Fe may not be so severe.

More accurate calculation of the electron capture rates for the NSE composition might also be important; if the rates by Mazurek et al. (1974) are employed for NSE, the total rate is smaller by a factor of ~ 2, which significantly reduces the ^{54}Fe/^{56}Fe ratio (Thielemann, Nomoto, and Yokoi 1984).

To initiate a carbon deflagration at lower ρ_c than C6 would require an accretion as rapid as $\dot{M} > 10^{-6}\ M_\odot$ yr^{-1}, which may occur in the double white dwarf scenario for SN I progenitors. However, it must be noted that rapid accretion does not necessarily lead to lower ignition density (§ 13.2.2).

13.5.2 Intermediate Mass Elements, Si–Ca

The abundances of Si–Ca are important for comparison with the early time spectra of SN I (Branch 1983). Both models (W7 and C6) eject ^{40}Ca with the abundance ratio of $<^{40}$Ca/^{56}Fe$> \sim 1$, and Si, S, and Ar with the ratio of $<X_i/^{56}$Fe$> \sim 0.4$–0.5. Moreover, the relative abundances of Si, S, and Ar are remarkably close to the solar values. On the other hand, the abundances of odd Z elements in the Si–Ti range, i.e., P, Cl, K, and Sc, are as small as $<X_i/^{56}$Fe$> \sim 0.03$–0.07.

It is noteworthy that Si-peak elements are synthesized in the layer with relatively high density, i.e., $\rho_{p,7} = 2.5$–4. This is significantly higher than the 10^5–10^6 g cm^{-3} in the oxygen-rich layer of a 25 $M_\odot$ star where Si-peak elements are produced (Woosley and Weaver 1982). Such a difference may cause the difference in the abundance ratios among Si-peak elements.

13.5.3 s- and r-Process Elements

The carbon deflagration supernova is a possible source of s- and r-process elements. During the accretion onto the white dwarf, helium shell flashes recur many times when enough helium is accumulated below the hydrogen-burning shell. In such a case, the helium shell flash develops a convective zone which mixes hydrogen into the helium layer. Subsequent reactions of ^{12}C$(p,\gamma)^{13}$N$(e^+\nu)^{13}$C$(\alpha,n)^{16}$O could produce neutrons to synthesize s-process elements. Even without such a mixing of hydrogen, the peak temperature during the helium shell flashes is high enough to produce neutrons via

^{22}Ne$(\alpha,n)^{25}$Mg reaction for $M \gtrsim 1\ M_\odot$ (Iben 1975, 1981; Sugimoto and Nomoto 1975). Therefore, if hydrogen or helium accretes onto the white dwarf, the outer layer of the white dwarf contains some s-process elements at the time of explosion.

The carbon deflagration wave produces a precursor shock wave as discussed in § 13.3.1. When the deflagration wave reaches the He layer near the white dwarf surface, the precursor shock grows strong because of the steep density gradient. If the temperature at the passage of the shock wave becomes high enough for ^{22}Ne$(\alpha,n)^{25}$Mg reaction to generate neutrons, the r-process could operate on the seed s-process elements (Truran et al. 1978; Thielemann et al. 1979; Cowan et al. 1980). The amount of s- and r-process elements produced by this mechanism deserves further quantitative investigation.

It should be noted that these s- and r-processes do not operate for the CO-CO double white dwarf system since no He layer is formed.

13.5.4 γ-Radioactivities

Besides ^{56}Ni and ^{56}Co, SN I produces some γ-radioactivities; i.e., ^{22}Na$(2 \times 10^{-7}\ M_\odot)$, ^{26}Al$(4 \times 10^{-6}\ M_\odot)$, ^{44}Ti$(8 \times 10^{-5}\ M_\odot)$, ^{48}V$(3 \times 10^{-7}\ M_\odot)$, ^{60}Fe$(2 \times 10^{-9}\ M_\odot)$, and ^{57}Co$(2 \times 10^{-3}\ M_\odot)$ are ejected from the model W7. The ratio ^{26}Al/^{27}Al is 6×10^{-3}, which is larger than the ratio in the 25 $M_\odot$ model.

13.6 Contribution of SN I to Galactic Nucleosynthesis

If we adopt a carbon deflagration supernova model like case W7 as a standard SN I model, SN I make a significant contribution to galactic nucleosynthesis in the Si–Ni range.

It has been pointed out that the intermediate mass elements, Si–Ca, are underproduced relative to ^{16}O in the ejecta of a 25 $M_\odot$ star which may be a typical site of massive star nucleosynthesis (Woosley and Weaver 1982). Woosley and Weaver (1982) proposed that elements in the range of S–Ti could be produced by stars more massive than 25 $M_\odot$ or very massive population III stars. Arnett and Thielemann (1984) showed that some of these problems are removed with an enhanced ^{12}C$(\alpha,\gamma)^{16}$O rate in He burning.

Nomoto, Thielemann, and Wheeler (1984) have made another proposal: figure 13.13 shows the combined nucleosynthesis products from the W7 model

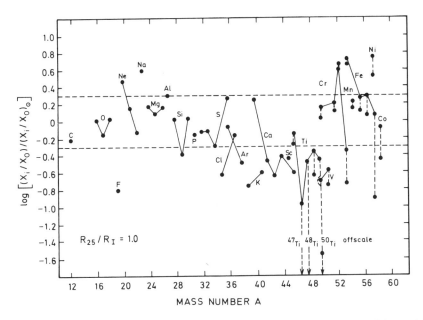

Fig. 13.13. The combined nucleosynthesis products from the model for SN I (W7) and the 25 $M_\odot$ star assuming equal explosion rates ($R_I = R_{25}$). The abundances of isotopes relative to ^{16}O are normalized to solar values. The vertical dashed lines connect to points for which the 25 $M_\odot$ star iron-peak elements are omitted. These elements are very sensitive to the mass cut in the core collapse.

of SN I and the 25 $M_\odot$ model, where the explosion rate of 25 $M_\odot$ stars, R_{25}, is assumed to be equal to that for SN I, R_I. This combination shows a rather good fit to the solar abundance ratios over a wide range in A. This suggests that a significant fraction of Si–Ca originate from SN I to fill the gap of SN II products; i.e., nucleosynthesis in the deflagration model for SN I is quite complementary to SN II.

The absolute rates of element production give some constraints on the SN rate. Nomoto, Thielemann, and Wheeler (1984) compared the Fe production rate from SN I with the observed rate in the solar neighborhood (Twarog and Wheeler 1982). Here we follow Tinsley (1980) and make a crude estimate of the abundances of elements, X_i, ejected from SN I by the yield $X_i \equiv m_i r/s$; here m_i is the mass of species i ejected by each SN I, r is the SN I rate, and s is the net star formation rate. If we adopt values of $s = 5$ $M_\odot$ pc^{-2} Gyr^{-1} (Tinsley 1980) and $r = (0.01-0.03)$ pc^{-2} Gyr^{-1} (Tammann 1982), $X_i = (0.002-0.006)$ $(m_i/M_\odot)$. Model W7 gives $X(Fe) = 0.0017-0.0051$, which is

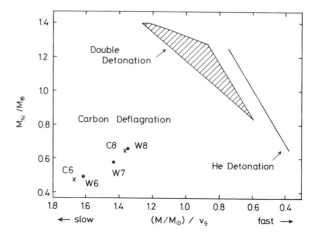

Fig. 13.14. The brightness ($\sim M_{Ni}$) and the speed class (indicated by M/v) of the light curves due to radioactive decays. Possible ranges for various types of explosion are indicated, i.e., He detonation in He white dwarfs, double detonations in C+O white dwarfs, and several carbon deflagration models. Here $v_9 \equiv v/10^9$ cm s^{-1}).

marginally consistent with the solar abundance if we adopt the lower limit for the SN I rate. If the SN I rate is higher, part of the Fe from SN I may form grains or escape from the Galaxy.

Assuming $r = 0.01$ pc^{-2} Gyr^{-1} which correspond to 10^{-2} yr^{-1} for the Galaxy (Tammann 1982), model W7 gives $X(^{56}Fe)/\odot \sim 0.91$, $X(^{40}Ca)/\odot \sim$ 1.2, $X(^{36}Ar)/\odot \sim 0.45$, $X(^{32}S)/\odot \sim 0.37$, $X(^{28}Si)/\odot \sim 0.42$, $X(^{24}Mg)/\odot \sim$ 0.080, $X(^{16}O)/\odot \sim 0.033$, and $X(^{12}C)/\odot \sim 0.017$, respectively, relative to the solar abundances. Therefore SN I could produce a significant fraction of ^{40}Ca and Si-peak elements in the Galaxy besides iron-peak elements.

13.7 Comparison with Observations

13.7.1 Light Curves

The carbon deflagration supernova models presented here eject 0.5–0.7 $M_\odot$ ^{56}Ni. This is enough to power the SN I light curve by the radioactive decays. The shape of theoretical light curves for this type of model shows a very good fit to the observations (Chevalier 1981; Arnett 1982; Sutherland and Wheeler 1984). Figure 13.14 shows a (M_{Ni}-M/v) plane which characterizes the light curves: the mass of ^{56}Ni, M_{Ni}, is approximately proportional to

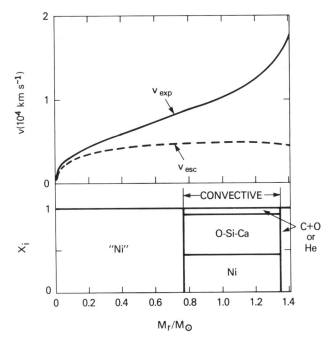

Fig. 13.15. Composition structure and velocity profile of the carbon deflagration model where the efficiency of convective mixing behind the deflagration wave was assumed to be extremely large.

the maximum luminosity and M/v is related to the width, i.e., speed class, of the light curves since the effective diffusion time is given by $\tau_m \equiv (\kappa M/v)^{1/2}$ (Arnett 1982) where M is the total ejected mass and $v \sim (E/M)^{1/2}$ is the velocity scale. Compared with the detonation-type models in figure 13.14, the light curves of the deflagration models are dimmer and broader, i.e., decline more slowly.

13.7.2 Early Time Spectra

The early time spectra of SN I provide information about the composition structure in the outer layer which is expanding at $\sim 10^4$ km s^{-1} (Branch 1983). The synthetic spectra for the model W7 have recently been calculated (Branch 1984). The fit to the observed spectra of SN I 1981b at maximum light and 17 days later is fairly good if the outer layers at $M_r > 0.7\ M_\odot$ are mixed. On the other hand, the spectra for the models with complete composition stratification do not show a good fit.

Although a mixing process after the passage of the deflagration wave was not included in model W7, convective mixing in the outer layers is expected to occur because of the decaying nature of the deflagration wave: the incinerated NSE region is almost isothermal and thus convectively stable. On the other hand, material in the outer layers at $M_r \gtrsim 0.7\ M_\odot$ is processed by the decaying deflagration wave and so the nuclear energy release gets smaller as the deflagration wave decays. Therefore, such layers are convectively unstable because entropy decreases outward to yield even a density inversion.

Such a mixing behind the deflagration wave was taken into account in the model by Nomoto (1980) where the convectively unstable layers were assumed to mix quickly. The resultant composition and velocity profile are shown in figure 13.15. More accurate modeling of such a mixing deserves further study.

13.7.3 Remnant Star

The carbon deflagration model predicts complete disruption of the star leaving no neutron star remnant behind. This is consistent with the failure to detect point X-ray sources or X-ray nebulae in the SN I remnants (Helfand 1983) and the latest model of neutron star cooling (Nomoto and Tsuruta 1981).

13.8 Concluding Remarks

13.8.1 Carbon Deflagration Models

Many of the observational features of SN I, in particular near maximum light, can be accounted for by the carbon deflagration model of accreting white dwarfs. Detailed nucleosynthesis calculation has shown that the carbon deflagration wave synthesizes $\sim 0.6\ M_\odot$ ^{56}Ni which is enough to power the light curve. In the outer layers of the white dwarf, substantial amounts of intermediate mass elements, Ca, Ar, S, Si, Mg, and O, are synthesized. The synthetic spectra based on this model show a good fit to the early time spectra of SN I. Such nucleosynthesis in SN I is complementary to the nuclear products from massive stars.

It must be remembered that the carbon deflagration models include a highly uncertain parameter, $\alpha = l/H_p$, and the results are rather sensitive to α. For smaller α, the explosion energy is too small (Nomoto et al. 1976), while a deflagration with $\alpha \sim 1$ produces almost exclusively iron-peak elements.

The excess neutronization enhances the ^{54}Fe abundance relative to ^{56}Fe by a factor of ~ 4 in the present models. This may give some restrictions on $\rho_{c,ig}$, α, $\dot{M}$, etc., in the models. The convective Urca processes would delay the occurrence of the thermonuclear runaway to higher density than $\rho_{c,ig}$ by a factor of ~ 1.5 (Iben 1982), which makes the restriction on $\rho_{c,ig}$ tighter. Further study may be needed for the evolution of the possible progenitor of SN I.

13.8.2 Detonation Models and Progenitors

Theoretical models of accreting white dwarfs predict the occurrence of another type of explosion for the slow accretion, namely, detonation supernovae initiated by a helium flash. However, the detonation-type explosion produces almost exclusively iron-peak elements, which is inconsistent with the observations.

Then the question is why the carbon deflagration occurs much more frequently in nature than the detonation-type supernova (Branch 1983). Possible answers are as follows:

1. For C+O white dwarfs, the carbon deflagration is simply more frequent than the double detonations because the deflagration occurs for more rapid accretion (see figure 13.2).

2. The occurrence of a detonation-type explosion is rare because hydrogen shell flashes for the slow accretion are so strong that most of the accreted matter is lost from the white dwarf and cannot build up a helium layer with sufficient mass (figure 13.1).

3. Most SN I originate from the double white dwarf system (Iben and Tutukov 1984; Webbink 1984). In this case, the accretion rate is so high that the double detonation supernova corresponding to slow $\dot{M}$ must be a rare event. The He-He white dwarf pair may not explode for some reason, e.g., too small a total mass. In this regard, the pre-maximum spectra of SN I 1983 in NGC 5236 show an interesting narrow emission feature which is unshifted and could be identified as a He I line (Richtler and Sadler 1983). This may suggest the existence of a (He-rich) nebula which could form during a rapid accretion in the double white dwarf system.

13.8.3 Collapse of White Dwarfs in Silent Supernovae

The thermonuclear explosion disrupts the white dwarf completely leaving no neutron star behind except for the case of a single detonation leaving a white dwarf remnant. On the other hand, the evolutionary origin of low mass X-ray binary systems and a binary containing the 6.1 ms pulsar might be the collapse of a certain class of accreting white dwarfs to form neutron stars (e.g., van den Heuvel 1981; Helfand et al. 1983). In this regard, possible effects of carbon and oxygen separation in the crystallizing white dwarf have been investigated by Isern et al. (1983) and Mochkovitch (1983) although such a separation is quite hypothetical (S. Ichimaru, private communication). If the separation occurs, a solid oxygen core is formed interior to the C+O layers. The most probable outcome of accretion onto such a white dwarf is off-center carbon ignition prior to the onset of electron captures on ^{16}O at the center because of the rather low ignition density of carbon ($\sim 4 \times 10^9$ g cm^{-3}). Therefore the outcome could be SN I rather than collapse to form a neutron star.

However, collapse of accreting white dwarfs can occur for O+Ne+Mg white dwarfs because electron captures on ^{24}Mg and ^{20}Ne trigger the collapse prior to the explosion (Nomoto et al. 1979; Miyaji et al. 1980). It must be noted here that the neon deflagration prior to the electron captures (Woosley et al. 1984) is unlikely; the reason for this is that the photodisintegration rate needs to be enhanced by strong screening in order to initiate a neon flash at such low T_c as $\sim 3 \times 10^8$ K, yet the screening effect cannot enhance the ^{20}Ne photodisintegration rate (see § 5 in Nomoto 1984c).

Since O+Ne+Mg white dwarfs form from 8–10 $M_\odot$ stars in close binary systems (Nomoto 1984a), the frequency of white dwarf collapse is several orders of magnitude smaller than SN I. It may be consistent with the statistics of low mass X-ray binaries and binary pulsars (van den Heuvel 1981; Webbink et al. 1983; Iben and Tutukov 1984). The collapsing white dwarf will make a bounce and eject some materials as found in 8–10 $M_\odot$ stars (Hillebrandt, Nomoto, and Wolff 1984). Because of lack of either ^{56}Ni or an extended envelope, however, the explosion would be *silent*.

Acknowledgments

I would like to thank Drs. F.-K. Thielemann, K. Yokoi, D. Branch, J. C. Wheeler, and I. Iben for the collaborative work on the carbon deflagration models, synthetic spectra, galactic nucleosynthesis, and double white dwarf system. I am also indebted to Drs. R. Kippenhahn, W. Hillebrandt, E. Müller, I. Iben, and J. W. Truran for stimulating discussion and hospitality during my stay at the Max-Planck-Institut and the University of Illinois, and to Prof. D. Sugimoto for useful comments and discussion. This work is supported in part by the Space Data Analysis Center of the Institute of Space and Astronautical Sciences, Tokyo, and by the Japanese Ministry of Education, Science, and Culture through research grant 58340023.

References

Arnett, W. D. 1969, *Ap. Space Sci.*, **5**, 180.

——. 1979, *Ap. J. (Letters)*, **230**, L37.

——. 1982, *Ap. J.*, **253**, 785.

Arnett, W. D., and Thielemann, F.-K. 1984, in *Stellar Nucleosynthesis*, ed. C. Chiosi and A. Renzini (Dordrecht: Reidel).

Arnett, W. D., Truran, J. W., and Woosley, S. E. 1971, *Ap. J.*, **165**, 87.

Axelrod, T. S. 1980a, in *Type I Supernovae*, ed. J. C. Wheeler, p. 80 (Austin: University of Texas).

——. 1980b, Ph.D. thesis, University of California at Santa Cruz.

Becker, R. H., Holt, S. S., Smith, B. W., White, N. E., Boldt, E. A., Mushotzky, R. F., and Serlemitos, P. J. 1980, *Ap. J. (Letters)*, **235**, L5.

Branch, D. 1983, *Ann. N.Y. Acad. Sci.;* review paper presented at the XI Texas Symposium on Relativistic Astrophysics.

——. 1984, this volume.

Branch, D., Buta, R., Falk, S. W., McCall, M. L., Sutherland, P. G., Uomoto, A., Wheeler, J. C., and Wills, B. J. 1982, *Ap. J. (Letters)*, **252**, L61.

Branch, D., Lacy, C. H., McCall, M. L., Sutherland, P. G., Uomoto, A., Wheeler, J. C., and Wills, B. J. 1983, *Ap. J.*, **270**, 123.

Bruenn, S. W. 1971, *Ap. J.*, **168**, 203.

Buchler, J. R., and Mazurek, T. J. 1975, *Mem. Soc. Roy. Sci. Liège*, **8**, 435.

Canal, R., and Schatzman, E. 1976, *Astr. Ap.*, **46**, 229.

Chevalier, R. A. 1981, *Ap. J.*, **246**, 267.

Colgate, S. A., Petschek, A. G., and Kriese, J. T. 1980, *Ap. J. (Letters)*, **237,** L81.

Courant, R., and Friedrichs, K. O. 1948, *Supersonic Flow and Shock Waves* (New York: Interscience).

Cowan, J. J., Cameron, A. G. W., and Truran, J. W. 1980, *Ap. J.*, **241,** 1090.

Duncan, M. J., Mazurek, T. J., Snell, R. L., and Wheeler, J. C. 1976, *Ap. Letters*, **17,** 19.

Ergma, E. V., and Tutukov, A. V. 1976, *Acta Astr.*, **26,** 69.

Fujimoto, M. Y., and Sugimoto, D. 1982, *Ap. J.*, **257,** 291.

Fuller, G. M., Fowler, W. A., and Newman, M. 1982, *Ap. J.*, **252,** 715.

Helfand, D. J. 1983, in *IAU Symposium 101, Supernova Remnants and Their X-Ray Emission,* ed. J. Danziger and P. Gorenstein, p. 471 (Dordrecht: Reidel).

Helfand, D. J., Ruderman, M. A., and Shaham, J. 1983, *Nature*, **304,** 423.

Hillebrandt, W., Nomoto, K., and Wolff, G. W. 1984, *Astr. Ap.,* in press.

Hoyle, F., and Fowler, W. A. 1960, *Ap. J.*, **132,** 565.

Iben, I., Jr. 1975, *Ap. J.*, **196,** 525.

——. 1981, *Ap. J.*, **243,** 987.

——. 1982, *Ap. J.*, **259,** 244.

Iben, I., Jr., and Tutukov, A. V. 1984, *Ap. J. Suppl.*, **54,** 335.

Ichimaru, S., and Utsumi, K. 1983, *Ap. J. (Letters)*, **269,** L51.

Isern, J., Labay, J., Hernanz, M., and Canal, R. 1983, *Ap. J.*, **273,** 320.

Ivanova, L. N., Imshennik, V. S., and Chechetkin, V. M. 1974, *Ap. Space Sci.*, **31,** 497.

Mazurek, T. J., Meier, D. L., and Wheeler, J. C. 1977, *Ap. J.*, **213,** 518.

Mazurek, T. J., Truran, J. W., and Cameron, A. G. W. 1974, *Ap. Space Sci.*, **27,** 261.

Meyerott, R. E. 1980, *Ap. J.*, **239,** 257.

Miyaji, S., Nomoto, K., Yokoi, K., and Sugimoto, D. 1980, *Pub. Astr. Soc. Japan*, **32,** 303.

Mochkovitch, R. 1983, *Astr. Ap.*, **122,** 212.

Müller, E., and Arnett, W. D. 1982, *Ap. J. (Letters)*, **261,** L107.

Nomoto, K. 1980, in *Type I Supernovae,* ed. J. C. Wheeler, p. 164 (Austin: University of Texas).

——. 1981, in *IAU Symposium 93, Fundamental Problems in the Theory of Stellar Evolution,* ed. D. Sugimoto, D. Q. Lamb, and D. N. Schramm, p. 295 (Dordrecht: Reidel).

——. 1982a, *Ap. J.,* **253,** 798.

——. 1982b, *Ap. J.,* **257,** 780.

——. 1984a, *Ap. J.,* **277,** 791.

——. 1984b, in *Stellar Nucleosynthesis,* ed. C. Chiosi and A. Renzini, p. 205 (Dordrecht: Reidel).

——. 1984c, in *Proc. Toulouse Workshop on Collapse and Numerical Relativity* (Dordrecht: Reidel).

Nomoto, K., and Iben, I. 1984, in preparation.

Nomoto, K., Miyaji, S., Sugimoto, D., and Yokoi, K. 1979, in *IAU Colloquium 53, White Dwarfs and Variable Degenerate Stars,* ed. H. M. Van Horn and V. Weidmann, p. 56 (Rochester: University of Rochester).

Nomoto, K., and Sugimoto, D. 1977, *Pub. Astr. Soc. Japan,* **29,** 765.

Nomoto, K., Sugimoto, D., and Neo, S. 1976, *Ap. Space Sci.,* **39,** L37.

Nomoto, K., Thielemann, F.-K., and Wheeler, J. C. 1984, *Ap. J. (Letters),* **279,** L23.

Nomoto, K., Thielemann, F.-K., and Yokoi, K. 1984, *Ap. J.,* submitted.

Nomoto, K., and Tsuruta, S. 1981, *Ap. J. (Letters),* **250,** L19.

Richtler, T., and Sadler, E. B. 1983, *Astr. Ap.,* **128,** L3.

Sugimoto, D., and Miyaji, S. 1981, in *IAU Symposium 93, Fundamental Problems in the Theory of Stellar Evolution,* ed. D. Sugimoto, D. Q. Lamb, and D. N. Schramm, p. 191 (Dordrecht: Reidel).

Sugimoto, D., and Nomoto, K. 1975, *Pub. Astr. Soc. Japan,* **27,** 197.

——. 1980, *Space Sci. Rev.,* **25,** 155.

Sutherland, P. and Wheeler, J. C. 1984, *Ap. J.,* **280,** in press.

Taam, R. E. 1980, *Ap. J.,* **237,** 142.

Tammann, G. A. 1982, in *Supernovae: A Survey of Current Research,* ed. M. J. Rees and R. J. Stoneham, p. 371 (Dordrecht: Reidel).

Thielemann, F.-K., Arnould, M., and Hillebrandt, W. 1979, *Astr. Ap.,* **74,** 175.

Thielemann, F.-K., Nomoto, K., and Yokoi, K. 1984, in preparation.

Tinsley, B. M. 1980, in *Type I Supernovae,* ed. J. C. Wheeler, p. 196 (Austin: University of Texas).

Trimble, V. 1982, *Rev. Mod. Phys.*, **54,** 1183.

Truran, J. W., Cowan, J. J., and Cameron, A. G. W. 1978, *Ap. J. (Letters)*, **222,** L63.

Twarog, B. A., and Wheeler, J. C. 1982, *Ap. J.*, **261,** 636.

Unno, W. 1967, *Pub. Astr. Soc. Japan*, **19,** 140.

van den Heuvel, E. P. J. 1981, in *IAU Symposium 93, Fundamental Problems in the Theory of Stellar Evolution*, ed. D. Sugimoto, D. Q. Lamb, and D. N. Schramm, p. 155 (Dordrecht: Reidel).

Weaver, T. A., Axelrod, T. S., and Woosley, S. E. 1980, in *Type I Supernovae*, ed. J. C. Wheeler, p. 113 (Austin: University of Texas).

Webbink, R. F. 1984, *Ap. J.*, **277,** 355.

Webbink, R. F., Rapaport, S., and Savonije, G. J. 1983, *Ap. J.*, **270,** 678.

Wheeler, J. C. 1982, in *Supernovae: A Survey of Current Research*, ed. M. J. Rees and R. J. Stoneham, p. 167 (Dordrecht: Reidel).

Woosley, S. E., Axelrod, T. S., and Weaver, T. A. 1984, in *Stellar Nucleosynthesis*, ed. C. Chiosi and A. Renzini (Dordrecht: Reidel).

Woosley, S. E., and Weaver, T. A. 1982, in *Essays in Nuclear Astrophysics*, ed. C. A. Barnes, D. D. Clayton, and D. N. Schramm, p. 377 (Cambridge University Press).

Woosley, S. E., Weaver, T. A., and Taam, R. E. 1980, in *Type I Supernovae*, ed. J. C. Wheeler, p. 96 (Austin: University of Texas).

Wu, C.-C.; Leventhal, M.; Sarazin, C. L.; and Gull, T. R. 1983, *Ap. J. (Letters)*, **269,** L5.

14

Carbon Combustion Models of Type I Supernovae: The Propagation of the Thermonuclear Burning Front

Ewald Müller and W. David Arnett

Abstract

To study the propagation of thermonuclear burning fronts in degenerate carbon-oxygen cores we have performed a set of numerical experiments which incorporate the treatment of two-dimensional (axisymmetric) convective flow without the need of a phenomenological theory of convection. Our numerical experiments show a variety of possible dynamical evolutions ranging from a spherical detonation to a nonspherical deflagration of the carbon-oxygen core. The major results are: (a) Comparison of numerical experiments having the same initial conditions but different spatial grid-resolution indicate that the nuclear burning in the star will probably propagate in form of a turbulent, inhomogeneous but on the average nearly spherical combustion front. (b) In all experiments except the one leading to a detonation, some carbon is not processed up to nickel. Typically 0.8–1.0 $M_\odot$ of nickel is produced in these experiments, resulting in light curves and spectra like those observed in Type I supernovae. (c) All experiments lead to a total disruption of the star, consistent with the absence of a compact object in the Tycho and Kepler supernova remnants. (d) The numerical experiments indicate that reliable nucleosynthesis predictions require more than one-dimensional models.

14.1 Introduction

Concerning the progenitors of Type I supernovae, the observational and theoretical situation is still controversial. As Type I supernovae are observed in elliptical and irregular galaxies as well as in the disk population of spirals, their progenitors are not unambiguously identifiable (see review by Branch

1982). Recent studies of white dwarfs in young galactic clusters seem to indicate that all single stars up to about 7 $M_\odot$ will become planetary nebulae and
white dwarfs (Reimers and Koester 1982; Weidemann and Koester 1983; see,
however, Wood et al. 1982). On the other hand it is argued that all single
stars with $M > 8\,M_\odot$ will eventually evolve toward the core collapse stage
and lead to a Type II supernova explosion (for a review see Hillebrandt 1982),
and progenitors of Type I supernovae may be stars in the mass range 6
$< M/M_\odot < 8$. A second class of progenitors are accreting white dwarfs in
binary systems, where, depending on the mass accretion rate, i.e., on the complicated details of the binary evolution, the mass transfer from a close companion leads to the explosion of the white dwarf (Nomoto 1982a, b; Iben and
Tutukov 1983).

The question whether under conditions of high electron degeneracy the
$^{12}C + {}^{12}C$ ignition would develop into an explosion and, in particular, a detonation of the star, leaving no dense remnant, and produce sufficient ^{56}Ni to
power the light curve of a Type I supernova (in the manner suggested by Colgate and McKee 1969 for stars undergoing gravitational collapse) has been a
matter of controversy ever since Arnett (1968, 1969a) suggested this kind of
evolutionary behavior.

Due to a number of new developments much of the original proposal of
Arnett (1968, 1969a) still seems viable.

i) The situation regarding "overabundances" of Fe in supernovae and
supernova remnants has changed drastically (see review by Trimble 1983).
The late time spectra of Type I supernovae imply excess Fe (Kirshner and
Oke 1975) of order 0.5 $M_\odot$ (Meyerott 1980). The non-LTE excitation is best
explained by the decay of ^{56}Ni and ^{56}Co (Meyerott 1980; Axelrod 1980), and
the light curves are now well understood in terms of these decays (Trimble
1983; Arnett 1982). Recently, strong Fe^+ resonance absorption lines have
been discovered in the continuum of the IUE spectrum of the dwarf star pro-
jected near SNR 1006, and interpreted as a mass of $0.03 < M(Fe)/M_\odot < 1$ of
Fe (Wu et al. 1983).

ii) Several years ago the yield of Fe from the thermonuclear explosion of
intermediate mass stars seemed to be inconsistent with observations interpreted on the basis of galactic evolutionary models (Arnett 1974; Ostriker,
Richstone, and Thuan 1974). The recent work of Miller and Scalo (1979),
which points to a nearly constant stellar birthrate over galactic history,

together with newer estimates of the nickel (and eventually Fe) yield of less than 1.4 $M_\odot$ per event (see below), greatly weakens the Fe yield argument.

iii) In spite of the enormous amount of data collected on pulsars (Taylor and Manchester 1977; Lyne 1982), our theoretical understanding of the ages and formation of the objects is still crude, as the surprise resulting from the slowing down rate measured for the fast pulsars 1937+214 (Backer et al. 1982) and 1953+29 (Boriakoff et al. 1983) proves. The suggestion that pulsars are formed from intermediate mass stars (Gunn and Ostriker 1970) is therefore based on a poorly understood theory, and the alternative suggestion (Arnett 1969 b), that pulsars are derived from more massive stars, still seems viable.

iv) The result of Arnett (1969 a) that the carbon ignition does grow into an explosion has been the subject of much criticism (see reviews by Buchler, Mazurek, and Truran 1974; Mazurek and Wheeler 1980). Some of this criticism was based, however, on calculations which are themselves questionable. The "pulsational" results of Ivanova, Imshennik, and Chechotkin (1974), Buchler and Mazurek (1975), and Kudryashov, Shcherbatyuk, and Ergma (1979) are strongly influenced by neglect of the fact that the thermal instability occurs in a hydrodynamically unstable medium, where nonradial motions (convection) are already well developed.

A misconception which has had an effect on subsequent work is that "deflagration" calculations (see Sugimoto and Nomoto 1980 and references therein) include convection and thus remove an artificial aspect of the "detonation" calculation (Arnett 1969 a). As may be seen in figure 1 of Arnett (1968) and figure 10 of Arnett (1969 a), the evolution was followed from the initial stage of carbon ignition through the onset of convection and thermal runaway. The mass zones flashed individually, giving rise to steep temperature variations. The mixing-length approximation requires that the variation of quantities be slow over lengths shorter than the mixing length; the opposite was true. Both the "detonation" and "deflagration" calculations represent a gross oversimplification of the real situation. We have a thermonuclearly driven hydrodynamic instability growing from a nonspherical hydrodynamic flow pattern which is turbulent. Parameterizing the effective velocity of the flame front was a useful initial expedient, but now confuses our thinking about the physics of the process. As we shall see, the "detonation model" results are a correct representation of the physics for spherical symmetry; the model's weakness is not an artificial initiation of detonation, but rather a neglect of nonspherical effects, a flaw shared with the "deflagration model" results.

This neglect of the convective stage of evolution also affects the conclusions about detonation that were based on shock tube arguments (Mazurek, Meier, and Wheeler 1977). These authors did not really address the question of whether the detonation does or does not form in the conditions actually encountered in the evolutionary calculations, namely, in a strongly driven, convective, turbulent core. Their assumption that the flash would be followed by a quiescent phase has not been justified.

First, we shall describe the initial models we have adopted in our study (§ 14.2) and the input physics as well as the numerical method we have applied in the calculations (§ 14.3). The results of our numerical experiments are given in § 14.4 and analyzed in detail in § 14.5, with the emphasis of the discussion put on the propagation of the burning front.

14.2 Initial Models

The obvious way to study the evolution of the carbon core would be to do a multidimensional evolutionary calculation (to incorporate convection!) starting at the point of instability and following it through the runaway stage. At the runaway point ($T \approx 10^9$ K) the characteristic time for the temperature increase due to nuclear reactions is equal to the adiabatic sound travel time. But because of the slow rates of nuclear reactions at temperatures $T < T_{\text{runaway}}$, the carbon flash develops on time scales much longer than the dynamical time scale, thereby preventing a study of this "nondynamical" stage of the evolution with an explicit hydrodynamic code. For stability reasons, explicit calculations are restricted to timesteps dictated by the well-known Courant-Friedrichs-Lewy condition (e.g., Roache 1976).

Therefore we have decided to start with a study of the evolution of the carbon core beginning at the runaway stage. Clearly this is a compromise. The initial models for our calculations are ambiguous insofar as we have to make assumptions about the evolution between the point of ignition and the runaway stage. The "nondynamic" nuclear burning will influence mainly the temperature distribution in the central parts of the core, whereas the structure of the core, i.e., the density distribution, is only slightly affected due to weak dependence of the pressure on the temperature. To survey the evolutionary paths to the runaway stage, we have examined the evolution of a

Table 14.1
Properties of Initial Models

Property	Model			
	A	B	C	C′
$M_{\text{runaway}}/M_\odot$	5×10^{-3}	5×10^{-5}	5×10^{-3}	5×10^{-3}
Angular dependence of initial temperature distribution	No	No	Yes	Yes

Note: M_{runaway} = amount of mass with increased (runaway) temperature in units of a solar mass.

variety of initial models differing only in the temperature distribution in the core.

The common features of the initial models we have studied are as follows: All initial models are $n = 3$ polytropes with a central density of $\rho_c = 3 \times 10^9$ g cm^{-3} and a central temperature of $T_c = 4 \times 10^8$ K. The mass of this polytrope is 1.46 $M_\odot$ with an initial entropy per nucleon of 0.636 k_B, where k_B is the Boltzmann constant. For the evolutionary calculations and for the integration of the initial (spherically symmetric) hydrostatic equilibrium configurations we have used an equation of state consisting of a relativistic, degenerate electron gas including electron-positron pairs, a photon gas, and an ideal Boltzmann gas of mean molecular weight μ. All initial models had $\mu = 12$ throughout the whole core, i.e., a homogeneous pure carbon core.

This basic model was then modified. The temperature in a certain number of central mass zones was raised to a value of 2.8×10^9 K, at which immediate processing of carbon up to nickel takes place (see also § 14.3). The amount of mass which had an increased temperature was varied in the different models. Characteristics of the two initial models (A and B) are listed in table 14.1. In addition we have explored two models, C and C′, with a nonspherical initial temperature distribution, simulating probable inhomogeneous and anisotropic burning in the earlier phases of the evolution. Model C was obtained from model A by adding a small (radius dependent) quadrupole component to the temperature distribution of model A, leading to a hot blob on the symmetry axis. The blob was centered at a radius of 6×10^6 cm, and the temperature difference between "pole" and "equator" at that radius was about 7%.

14.3 Numerical Method and Model

To follow the hydrodynamic and thermonuclear evolution of the carbon core, we have integrated the equations of motion, the equation of mass conservation, equations describing the change of entropy and the consumption of the nuclear fuel together with the Poisson equation on a Eulerian grid using spherical polar coordinates and assuming axial and equatorial symmetry. This set of equations is solved with an explicit two-dimensional Eulerian hydrocode based on that of Müller and Hillebrandt (1981). Finite differences are used in radial direction and a Legendre expansion is used in angular direction. Two important changes have been made compared to the original version of the code (see also Müller and Arnett 1984):

i)   The time integration of the equations is done in two consecutive "steps," and

ii) Contrary to the code of Müller and Hillebrandt (1981) the entropy equation and the supplemented rate equation are differenced in *both* coordinate directions according to the donor-cell scheme (e.g., Roache 1976).

Because (1) we are primarily interested in the production of ^{56}Ni and (2) computer limitations in our exploratory calculations limited us to two species (^{12}C = "fuel," ^{56}Ni = "ash"), we used an effective nuclear reaction rate (including screening corrections). We scaled the rate in such a way that the ignition temperature is raised from $T = 10^9$ K (Arnett 1969a) to $T = 3.5 \times 10^9$ K, the latter temperature being the lower limit for burning carbon to nickel (Woosley, Arnett, and Clayton 1973). For the nuclear energy release we used a value of 7×10^{17} ergs g^{-1} which assumes that once carbon burning starts the material will be processed immediately to ^{56}Ni because of the resulting high temperatures ($T \geq 4 \times 10^9$ K), leading to a further liberation of 2×10^{17} ergs g^{-1} which has to be added to the pure ^{12}C + ^{12}C energy generation rate of 5×10^{17} ergs g^{-1}. The photodisintegration of ^{56}Ni, which acts as an energy sink, is not included in the code. The suppression of the rate at temperatures below 3×10^9 K diminishes the importance of the numerical errors due to artificial diffusion, because a larger amount of diffusion is required to raise the temperature in front of the combustion region

Table 14.2
Properties of Numerical Experiments

Property	Experiment						
	A	B1	B2	B3	C1	C2	C3
Initial model	A	B	B	B	C	C	C´
Spatial resolution of radial grid $(\times 10^6$ cm)*	2	2	1	0.5	2	1	1
Spatial resolution of angular grid (meshpoints)**	30	30	30	50	30	30	30

*This radial resolution was used inside a radius of 2×10^7 cm.
**The angular resolution in centimeters is dependent on the radial position!

from $T < 4 \times 10^8$ K to the (modified) runaway temperature $T = 3.5 \times 10^9$ K. Test runs with the unmodified carbon reaction rate gave grossly overestimated propagation speeds of the burning front as long as the grid was not refined a great deal (to reduce the numerical errors) which led to impractically small timesteps.

The calculations for experiments A, B1, B2, and C were done with 100 radial and 30 angular gridpoints or Legendre polynomials up to order 12. In the case of experiment B3 we used 120 radial and 50 angular gridpoints or Legendre polynomials up to order 20 (see table 14.2). In *none* of the calculations was an (explicit) artificial viscosity applied. A typical sequence consists of 15×10^3 timesteps, which translates into roughly 4 hours of CPU time on a Cray-1 computer.

14.4 Numerical Experiments

14.4.1 Experiment A

Numerical experiment A is best described as a spherical detonation. In model A the temperature of the innermost $5 \times 10^{-3} M_\odot$ of the core was raised to a value of 2.85×10^9 K, which is close to the (modified) runaway

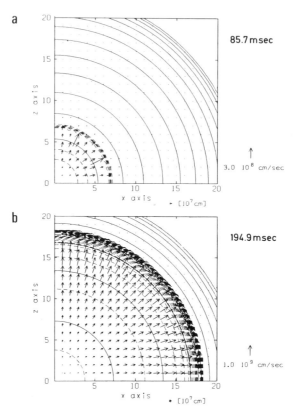

Fig. 14.1. Contour maps showing the evolution of the burning front for experiment A. The snapshots are taken at (a) 86 msec and (b) 195 msec. The solid lines are surfaces of constant density, the spacing being 0.5 on a logarithmic scale and the innermost contour corresponding to (a) log ρ = 9.0 and (b) log ρ = 8.5. The dashed lines are surfaces of constant carbon mass-fraction, the spacing being 0.1 on a linear scale; the mass-fraction is monotonically increasing outward. The length of the arrows is a measure of the velocity with the scale being defined by the length of the arrow and the numerical value in the lower right corner of each plot. The z-axis is identical with the axis of symmetry.

temperature. In this "hot" region roughly 6×10^{48} ergs of nuclear energy are liberated almost immediately ($t < 30$ msec), leading to the formation of a detonation wave.

Although this calculation was done with the full two-dimensional hydrocode, the deviations from spherical symmetry always remained negligible. In figure 14.1a one clearly recognizes the burning front corresponding to the steep entropy gradient (clustering of dashed contours). The second snapshot (figure 14.1b) is taken about 110 msec later in the evolution. The burning front has propagated to a radius of 1.8×10^8 cm; about 1.38 $M_{\odot}$ of fuel have been burned at that stage. The central density has dropped by a factor of 3.6 to a value of $\rho_c = 4.9 \times 10^8$ g cm^{-3}. The maximum velocity has increased from 3×10^8 cm s^{-1} to 10^9 cm s^{-1}.

The detonation front is moving with the local speed of sound relative to the matter immediately behind the front (the numerical errors being less than 10%), a behavior typical for a Chapman-Jouguet detonation (see Courant and Friedrichs 1948). The maximum velocity at the last model calculated ($t = 217$ msec) is 1.9×10^9 cm s^{-1} and the detonation has reached a radius of 2.15×10^8 cm, at about 90% of the core radius. At this time 1.41 $M_{\odot}$ of carbon have been burned and a total disruption of the core is inevitable.

14.4.2 Experiment B

Our next numerical experiment consists of a set of three calculations, namely, B1, B2, and B3. Their common feature is a "runaway" mass of 5×10^{-5} $M_{\odot}$ (model B), which is a factor 100 times smaller than in the case of experiment A.

14.4.2.1 *Experiment B1*

This experiment was done with the same angular and radial resolution as experiment A; i.e., experiments A and B1 differ only in "runaway" mass.

The early evolution ($t < 300$ msec) shows a very slowly propagating spherical burning front, the velocity being of the order 3×10^7 cm s^{-1}. Triggered at about 320 msec by the numerical noise present in all such calculations, a first deviation from sphericity happens in the convectively unstable combustion front near the symmetry axis. During further evolution this perturbation grows into a dominant finger-like feature. Parallel to the growth of the finger, the combustion front as a whole propagates outward with additional kinks appearing in the front.

After 650 msec the front (at the symmetry axis) has propagated out to $r = 1.8 \times 10^8$ cm and 0.55 $M_{\odot}$ of nickel has been produced. Between radii of 1.6×10^8 cm and 1.8×10^8 cm and near the axis, the flow is supersonic, with Mach numbers ranging up to 1.7. The shock, which is already slightly ahead of the burning region, is accelerated by the steep density gradient in the outer edge of the core. Close to the axis the heating due to shock compression is large enough to detonate the fuel.

14.4.2.2 *Experiment B2*

In this experiment the radial grid was refined relative to experiment B1, leaving the angular grid unchanged. By redistributing 100 radial gridpoints within a radius $r = 2 \times 10^7$ cm, the (equidistant) grid spacing was reduced from $r = 2 \times 10^6$ cm to $r = 1 \times 10^6$ cm.

In a previous publication (Müller and Arnett 1982) we have shown that with this kind of grid the burning front develops three finger-like features. Figure 14.2 displays the early evolution of the front which leads to the formation of the three "fingers." In the last snapshot the three kinks which will finally grow into the "fingers" are clearly visible. The velocities at 571 msec are supersonic between 1.2 $M_\odot \lesssim M(r) \leq 1.4$ $M_\odot$. We find that the region of maximum velocity (at $M(r) = 1.4$ $M_\odot$) is obviously ahead of the burning front $(M(r) \approx 1.1$ $M_\odot)$; i.e., the burning front stays behind the shock (see figure 2b in Müller and Arnett 1982). If the radial velocity is plotted versus radius, one finds $v(r) \sim r$ behind the shock, i.e., a homologous expansion.

The central density as a function of time exhibits an oscillatory behavior superimposed on a slow monotonic decrease for $t < 0.4$ s. For $t > 0.4$ s, we find a rapid drop of the central density to a value 10 times smaller than the initial value. This drop in density, the homologous expansion, and the liberated nuclear energy of 10^{51} ergs ($= 0.8$ $M_\odot$ of fuel consumed) strongly suggest a complete disruption of the core.

We extended our computations beyond the last model ($t = 571$ msec) of our previous publication (Müller and Arnett 1982) to see whether the shock will detonate the fuel in the steep density gradient near the core's edge. At $t = 590$ msec the shock front has come so close to the outer edge that we had to terminate the calculation because the numerical boundary conditions (fixed volume!) began to distort the flow. The shock front has reached a region where the density is about 10^6 g cm^{-3}, but up to this point there is no indication of a detonation of fuel.

14.4.2.3 *Experiment B3*

The third calculation within experiment B had both an increased radial and angular resolution compared to experiment B2. The (120!) radial gridpoints were distributed in such a manner that within $r = 2 \times 10^7$ cm the equidistant grid spacing was $\Delta r = 0.5 \times 10^6$ cm, a factor 2 better than in the previous experiment. The number of angular gridpoints was increased from 30 to 50, and the order of the highest polynomial used in the calculation was raised from 12 to 20.

The early evolution of the burning front is displayed in figure 14.3. The first deviation from sphericity occurs at a radius of 4×10^6 cm and at $t \approx 80$ msec. Note that for experiment B3 these deviations take first place at half the radius compared with experiment B2 (figure 14.2). On the other hand the time for the burning front to reach a radius of 8×10^6 cm is approximately the same for experiments B2 (131 msec) and B3 (119 msec). Between $t \approx 80$ msec and $t \approx 120$ msec more kinks appear in the burning front, resulting in five finger-like features (figure 14.3f).

At $t = 179$ msec the burning front (on average) has reached a radius of 1.4×10^7 cm, the shape and the relative and absolute size of the fingers being changed considerably (figure 14.4). There are two narrow fingers, one near the equator and the other at about $\theta = 45°$. The finger which has been situated between those two originally (figure 14.3f) has shrunk in relative (but not in absolute!) size and is now comparable to the other smaller kinks in the burning front. The upper two fingers (counted from the symmetry axis, only one is displayed in figure 14.4!) are broader than the first two but are of approximately the same length. In addition figure 14.4 shows several vortices which are set up by the burning.

Due to the fine zoning used in this experiment, the computational time was so large (see § 14.2) that we could not follow the evolution as far as in the case of experiment B2.

14.4.3 Experiment C

Experiment C consists of a set of three subexperiments. In all three cases the "runaway" mass was 5×10^{-3} solar masses (model C); unlike the former experiments the initial temperature distribution was nonspherical.

14.4.3.1 *Experiment C1*

This experiment was done with the same radial and angular resolution as experiments A and B1. Furthermore, experiments A and C1 had the same "runaway" mass (see table 14.2). The initial model of experiment C1 had a

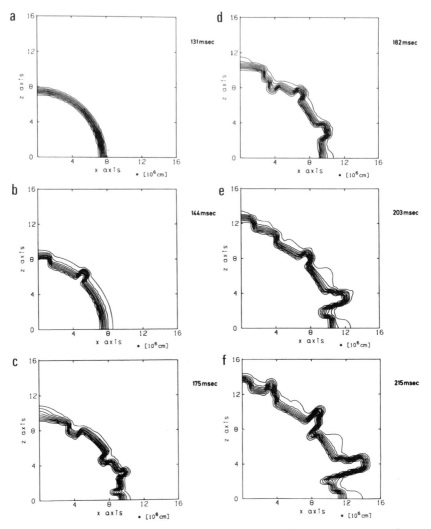

Fig. 14.2. Contour maps showing the early evolution of the burning front in experiment B2 starting from the top of the left column on down and continuing from the top of the right column on down. The contours are surfaces of constant carbon mass-fraction, the spacing being 0.1 on a linear scale. The mass-fraction is monotonically increasing outward. The axis of symmetry is identical with the z-axis.

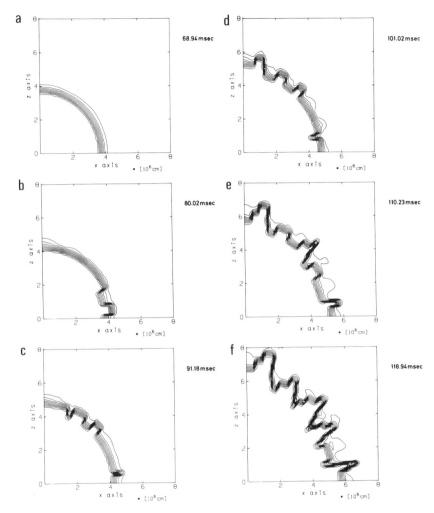

Fig. 14.3. Same as figure 14.2 but for experiment B3.

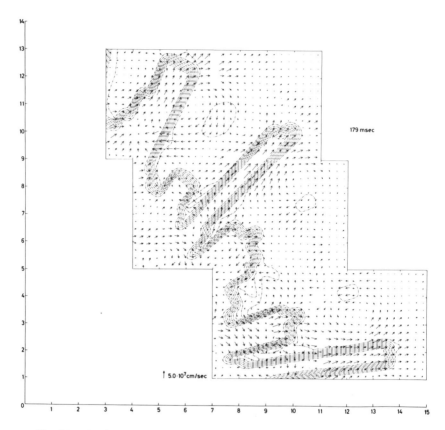

Fig. 14.4. Section of the burning front of experiment B3 at 179 msec. The contours are surfaces of constant carbon mass-fraction, the spacing being 0.1 on a linear scale.

nonspherical temperature distribution consisting of a hot blob at the axis of symmetry at a radius of 6×10^6 cm, the temperature contrast between the symmetry axis and the equator being about 7% (see § 14.2).

First, the fuel in the "blob" region is burned. After roughly 20 msec the burning front has reached the center of the star and has propagated outward along the symmetry axis to a radius of 1.7×10^7 cm. The further evolution of the burning front is dominated by the fact that the (radial) velocity of the front is smaller near the equator than at higher latitudes. This retardation of the burning near the equator is probably due to the symmetry imposed in the computational scheme.

Several finger-like features have been formed at 220 msec, the finger tips being situated at a radius of 1.5×10^8 cm. At the equator the front has only reached a radius of 4×10^7 cm. Up to this stage about 0.72 $M_\odot$ of carbon have been burned, resulting in a liberated energy of 10^{51} ergs. The central density has dropped to a value of 4.7×10^8 g cm^{-3}.

After 236 msec the flow has become supersonic in a narrow ($\Delta r < 2 \times 10^7$ cm), approximately spherical shell ahead of the burning front. When the shock reaches the steep density gradient at the core's edge, it is accelerated. This first happens near the axis of symmetry, where the strengthened shock eventually detonates the fuel.

14.4.3.2 *Experiment C2*

Experiment C2 is a repetition of experiment C1 with a finer radial grid, but with unchanged initial conditions. For experiment C2 a radial grid spacing of 1×10^6 cm instead of 2×10^6 cm was used inside a radius $r < 2 \times 10^7$ cm. Therefore the pair of experiments C1 and C2 corresponds to the pair B1 and B2 as far as the grid resolution is concerned. Figure 14.5 displays the evolution of the burning front. The main qualitative difference between C2 and C1 is that the burning front stays closer to the symmetry axis in case C2. The radius of a cylinder that contains the whole burning front at 220 msec is 7×10^7 cm in the case of experiment C2 and 10×10^7 cm that of experiment C1. Near the symmetry axis in both cases the front has reached a radius of roughly 15×10^7 cm and the flow pattern is very similar too. The different shapes of the burned volumes are reflected in the fuel consumption, which at 220 msec is 0.50 $M_\odot$ for experiment C2 and 0.72 $M_\odot$ for C1.

As in the previous experiment the flow finally becomes supersonic in a narrow region ahead of the burning front, which is less structured than in experiment C1. Up to this point ($t = 280$ msec) 0.64 $M_\odot$ of carbon have been consumed and the density has dropped to a value of 7.3×10^8 g cm^{-3}, strongly suggesting a complete disruption of the core.

14.4.3.3 *Experiment C3*

This experiment was done with the same radial grid as experiment C2, but the initial conditions were changed. Instead of a blob at the symmetry axis, the initial temperature distribution contained a nonspherical component,

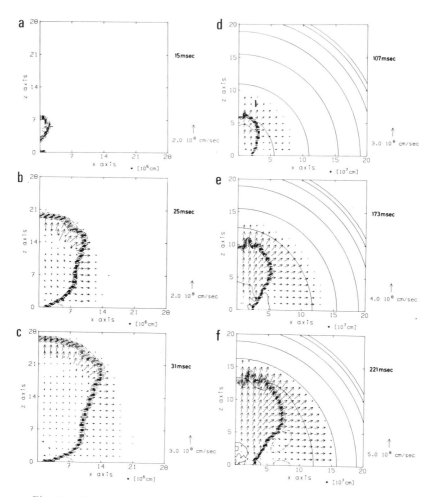

Fig. 14.5. Contour maps (and flow patterns) showing the evolution of the burning front of experiment C2. The spacing of the carbon mass-fraction contours (*dashed lines*) is 0.1 on a linear scale. The spacing of the density contours (*solid lines*) is 1.0 on a logarithmic scale, the outermost contour line corresponding to a density of 10^3 g cm^{-3}. Notice the change of scale between (*c*) and (*d*).

giving rise (due to axial symmetry) to a hot ring at the equator. The temperature distribution was obtained according to $T'(r,\theta) = T(r,90 - \theta)$, where $T(r,\theta)$ and $T'(r,\theta)$ are the initial temperature distribution of experiments C2 and C3, respectively.

The development of the burning within the first 100 msec is shown in figure 14.6. In the earliest snapshot (figure 14.6a) the toroidal burning front has just reached the center of the star. In the next 5 msec the size of the region containing burned material grows by more than a factor of 2, resulting in the oblate shape of the front. After 37 msec the fuel is depleted inside a radius of 3×10^7 cm and the shape of the front is that of a spherical shell (figure 14.6d). But the spherical appearance of the front is only a transient feature. Due to the initial conditions—a toroidal explosion—the material is accelerated by the driving pressure gradients faster in the direction parallel to the symmetry axis (converging flow!) than it is in the direction perpendicular to the axis; i.e., the front propagates faster along the symmetry axis than along the equator. After offsetting the original oblate conditions, which takes about 37 msec (figure 14.6d), the front gets a prolate shape (figures 14.6e and 14.6f).

At $t = 106$ msec (figure 14.6f) the burning front has reached a radius of 10×10^7 cm at the symmetry axis and 7×10^7 cm at the equator. About 0.78 $M_\odot$ of fuel have been burned, and the central density has dropped by a factor of 3. As in the two previous experiments, later in the evolution a shock front forms, but this time partially ahead and partially behind the burning front. We followed the evolution until $t = 213$ msec, when the shock front reaches the edge of the core. At that point 1.2 $M_\odot$ of nickel have been produced and the central density has dropped to 3×10^8 g cm^{-3}.

14.5 Discussion

14.5.1 Propagation of the Burning Front

The propagation of a subsonic combustion front is in general mediated by a diffusion process such as conduction or radiation and/or by convection. In the latter case the speed of the front can be quite different depending on the nature of the convective flow, i.e., whether the flow in question is large-scale convection or turbulent convection. Under the typical conditions in the core the ordinary diffusion can be neglected (Arnett 1969a).

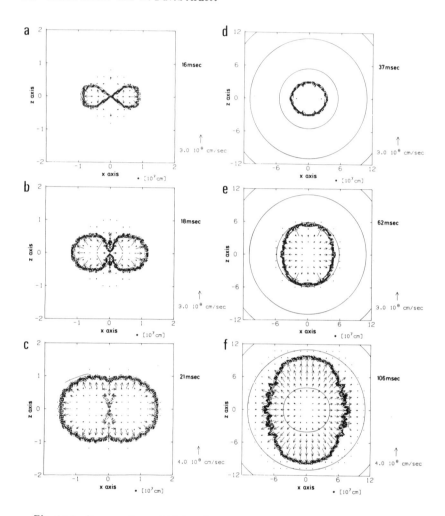

Fig. 14.6. Same as figure 14.5 but for experiment C3. Notice, however, that in this case the symmetry axis is a fictitious vertical line through $x = 0$.

Let us now consider the propagation of a burning front on the finite Eulerian grid used in our calculation. Without loss of generality we will restrict ourselves to a one-dimensional problem. Assume the front has reached a certain point in space coinciding with a cell boundary of the numerical grid. The hot matter of the burned cell expands into the unburned colder cell ahead of the front; i.e., there is a flow of hot matter into the cold cell. The advected hot upstream matter is completely mixed with the colder downstream matter. This complete mixing on the scale of the numerical grid is an inherent feature of all Eulerian calculations (without special interface methods). The mixing will lead to a temperature increase in the cell ahead. Finally the temperature becomes high enough to ignite the nuclear fuel: the cell flashes and the front advances one gridpoint. In addition the propagation of the burning front can be erroneously enhanced due to numerical diffusion, which inevitably accompanies the finite grid resolution of the calculation. The numerical diffusion, which may be orders of magnitude larger than any physical diffusion, can be reduced by using a less diffusive numerical scheme and/or a finer grid.

What follows from these general considerations for our calculations? First of all it is clear that we can only resolve structures in the burning front which have a size larger than the respective grid spacing. Comparing the experiments of set B, we see that the finer we made the grid the more (and smaller) structures the burning front exhibits. This strongly suggests that the initially spherical but Rayleigh-Taylor unstable burning front has a tendency to cascade to smaller and smaller scales. In the first two cases (B1 and B2) the cascading was limited by the grid resolution, which is demonstrated by doing finer zoned experiments, namely, B2 and B3, respectively. There is also a strong indication that the limiting physical scale is not yet reached in the finest zoned experiment B3, because the number of fingers seen (five) again reflects the minimum (angular) scale resolvable with Legendre polynomials of order 20.

Second, the general considerations above imply that the propagation speed of the front as given by the code might be either overestimated, because of the numerical diffusion present in the code, or underestimated, because the burning could be turbulent, i.e., the flow is not smooth on the scales resolved, thereby giving rise to a large effective diffusivity.

The interplay of advection, Eulerian mixing, numerical diffusion, and the finiteness of the grid causes the time differences found in the early evolution ($t < 100$ msec) of the experiments of set B. Whereas it takes about 52 msec to burn the "runaway" mass in B1, the same mass is burned within 29 msec in experiments B2 and B3. In experiment B1 the "runaway" mass is contained within one radial zone. The overpressure due to the increased temperature leads to an expansion of the matter in the first zone. With the flow also heat is advected into the next zone, reducing thereby the temperature in the first zone. This is counterbalanced by the nuclear energy production. The initial temperature of the innermost zone of 2.81×10^9 K should lead to the depletion of the fuel within 25 msec. The delay of about 27 msec in experiment B1 is due to advection and numerical diffusion, which reduce the heat content of the innermost zone. Notice that a 6% decrease in temperature is enough to change the fuel depletion time from 25 msec to 52 msec.

In experiments B2 and B3 the numerical diffusion is reduced due to the better zoning and the volume containing the "runaway" mass is radially resolved; i.e., there are only small pressure gradients between zone one and zone two but large gradients between zone two and zone three. Within the (original) fuel depletion time of 25 msec the temperature of the innermost zone is only weakly modified and the zone is depleted of fuel in 28 msec, corresponding to a 0.6% drop in temperature. The resulting overpressure and heat flow is large enough to flash the second zone 1 msec later.. A similar argument holds for experiment B3, where the "runaway" mass is also burned within 29 msec. This agreement between experiments B2 and B3 suggests that numerical diffusion in the better zoned models is not falsifying the results significantly, because otherwise there should be a time difference in depleting the "runaway" mass.

The propagation of the burning into the region with initially unmodified temperature (see § 14.2) also exhibits the effects of zone size. In experiment B1 a larger amount of matter must be heated up to the runaway temperature because only a *whole* cell can flash. On the other hand, more and more nuclear energy is liberated by burning more and more zones resulting in a faster advection; i.e., the time required to burn the next cell becomes smaller and differences due to zoning become less significant. The "disadvantage" of the coarser zoned experiments practically disappears when the ratio of the mass in the cell to be burned next to the mass of the already burned matter, which determines the advection speed, becomes smaller than 1. This is the case for experiment B1 at a radius of 8×10^6 cm. To reach this radius the burning front needs 270 msec in experiment B1, 131 msec in experiment B2, and 119 msec in experiment B3. The small differences between B2 and B3 once more confirm our earlier statement that diffusion errors in the finer zoned experiments are not very important. Notice that in experiment B2 the burning front is still spherical at 131 msec, whereas the front in experiment B3 already exhibits a lot of structures, which (as we shall show) are the reason for the different propagation speeds.

Besides the finite zone size we also have to take into account the varying zone shape, i.e., the two-dimensional effects. Whether we have radially elongated cells, which is the case near the center of the core for the spherical coordinate system we used, or more quadratic cells influences the propagation of the burning front to a certain degree.

Let us consider the following situation: At a certain radius R the burning front has propagated into the next radial (grid) shell only at one (arbitrary) angular grid point; i.e., the otherwise spherical front has a kink at that angular position. As the hot matter in this cell tries to expand in radial and angular direction there is a competition between flashing another radial cell or the neighboring angular cell. If the front "moves" faster in angular direction the kink will be smoothed out. Otherwise the kink gets even more pronounced and a finger grows.

The different spreading of the front away from the symmetry axis found in experiments C1 and C2 can be understood in this context. In experiment C2 the cells at a certain radial distance were less elongated than in experiment C1, which had a coarser radial grid but the same angular grid. Therefore in experiment C1 it took more time to flash the next cell in radial direction than in experiment C2. Consequently there was more time for the angular propagation of the front. Although this influence of the grid becomes unimportant later in the evolution (see arguments above), this modification of the early propagation remains imprinted on the front.

The competition between radial and angular propagation of the combustion front is the key to understanding at which stage the initially spherically symmetric fronts in experiment B start to develop "fingers." From the arguments just made it follows that the front will stay spherical until the angular advection and diffusion is slow enough for the radial (Rayleigh-Taylor) instability to develop. This happens after 300 msec in experiment B1, after 140 msec in experiment B2, and after 80 msec in experiment B3, the corresponding radii being 10×10^6 cm, 8×10^6 cm, and 4×10^6 cm. We see that while in experiments B1 and B2 the radius at which the instability starts to develop is roughly the same, it is a factor of 2 smaller in the case of B3. This reflects the fact that we have used the same angular grid in the first two experiments and a finer (by a factor 3/5) angular grid in experiment B3, and that (see above) the differences in the radial grids are of minor importance (at those radii) for the propagation of the combustion front.

The further evolution shows a qualitatively different behavior on the experiments of set B, again due to the grid resolution. Except for entropy and carbon mass fraction, all variables are expanded in angular direction in Legendre polynomials up to a certain order ι_{max}. Therefore the maximum angular resolution, which depends on radius, is given by $\Delta\theta = r/\iota_{max}$, where $\Delta\theta$ is the minimum angular scale resolvable and r the radius in question. If the maximum Legendre polynomial used is of order ι_{max}, one can only represent a function of θ with less than ι_{max} zeros in the range $0 < \theta < \pi$. In all experiments except experiment B3 we are able to resolve up to three finger-like features in the interval $0 < \theta < \pi/2$, and up to five in B3. Additional features in the burning front of smaller angular scales can show up to some extent (i.e., figure 14.4) as a consequence of the better resolved entropy and mass fraction distributions.

The behavior of experiment B1 is exceptional in the sense that it exhibits the influence of the imposed axial symmetry most strongly. In B1 the instability triggered by numerical noise first occurs at the symmetry axis. A coarser radial grid results in a larger perturbation; as the instability grows fastest at the axis, the front develops a dominant finger there. Therefore in experiment B1 (see also experiment C) the overall appearance of the burning front is not determined by the limiting angular scale (at large radii $[r > 10^7$ cm] the angular "grid" is coarser than the radial grid), but by the initial perturbation. In contrast to experiment B1, in experiments B2 and B3 the limiting angular scale is obviously responsible for the overall structure of the burning front: three fingers in experiment B2 (figure 14.2f) and five fingers in experiment B3 (figure 14.3f) reflect the order of the Legendre polynomial expansion. A comparison of the evolution of these three experiments leads to the conclusion that the front is trying to cascade to smaller and smaller scales and would probably, if not limited by a numerical resolution, become turbulent.

Although the finite grid is clearly a shortcoming of our approach, one can overcome this deficiency by comparing experiments having different spatial resolution, and extracting the (hopefully) grid independent results. The second shortcoming, the restriction to two-dimensional flow, may seem even more severe, especially when one remembers that turbulent convection in stars cannot be described by two-dimensional eddies, because turbulence is essentially a three-dimensional physical phenomenon. But as we are limited in spatial resolution to the computation of the relatively smooth flow of larger convective cells, and as the convectively (or Rayleigh-Taylor) unstable entropy gradient is in a radial direction, we think that our hydrodynamical approach is more appropriate than the mixing length approach. For the latter one must introduce a phenomenological length scale to study the dynamics, while in our approach the upward and downward meridional motions set up by the radial entropy gradient directly follow from the equations, which we solve. The neglected azimuthal fluid motions are probably less important for the qualitative features of the burning front than the meridional fluid motions, because the driving force—gravity—is in the radial direction.

The restriction to two-dimensional convection is a shortcoming of any numerical calculation capable of treating only two spatial dimensions. The restriction to axial symmetry gives rise to a more specific kind of shortcoming, namely, that different parts of the computational grid are topologically distinguishable. A kink in the burning front at the axis of symmetry or close to it almost resembles a blob-like perturbation of the front, but the same kink further off the axis is actually a ring-like or torus-like perturbation of the front. In other words a perturbation of the front near the axis of symmetry is a real three-dimensional perturbation; a perturbation elsewhere in the burning front is actually only a two-dimensional one. This topological difference gives rise to a angular dependent growth rate for perturbations of the burning front and thereby explains the small differences in the size of the fingers in experiment B2 and B3, and the exceptional behavior of experiment B1.

To understand this a bit better, let us consider a spherical blob of radius r and a (circular) torus with the same (internal) radius. For the blob, the ratio of surface to volume is given by the expression $3/r$, but the same ratio for the torus is given by the expression $0.5/r$. The blob has a larger surface relative to its volume when compared to a torus. Consequently a blob-like perturbation will grow faster than a torus-like perturbation because more fuel can be entrained per unit volume, if otherwise identical conditions are assumed.

Finally, the different behavior of experiments C1 and C2 on one hand and experiment C3 on the other is once more a direct consequence of the topologically different initial conditions (a point explosion and a ring explosion) in the two cases.

14.5.2 Implications

Several points stand out:

a) Only the detonation remained spherically symmetric. We expect the nuclear burning to propagate as an inhomogeneous, probably turbulent, combustion front.

b) Only the detonation processed all the fuel. Typically 0.8 to 1.0 $M_\odot$ of nickel was produced, which seems promising for models of Type I supernovae.

c) All experiments led to total disruption of the star, consistent with the apparent absence of a compact object in the Tycho and Kepler supernova remnants.

d) Because of the strong dependence of nucleosynthesis on temperature, the yields expected from temperature structures found here will seldom be accurately approximated by those obtained from spherically averaged temperature structures.

Acknowledgments

This work was supported in part by the NSF through grants AST 8022876 and 8316815 and by NASA through grant NAGW-123. One of us (E.M.) would like to express his gratitude to Professor Peter Meyer for the hospitality of the Enrico Fermi Institute of the University of Chicago, where part of this work was done.

References

Arnett, W. D. 1968, *Nature*, **219**, 1344.
——. 1969a, *Ap. Space Sci.*, **5**, 180.

——. 1969*b*, *Nature*, **222**, 359.

——. 1974, *Ap. J.*, **191**, 727.

——. 1982, *Ap. J.*, **253**, 785.

Axelrod, T. S. 1980, Ph.D. thesis, University of California at Santa Cruz, URCL-52994.

Backer, D. C.; Kulkarni, S. R.; Heiles, C.; Davis, M. M.; and Goss, W. M. 1982, *Nature*, **300**, 615.

Boriakoff, V., Buccheri, R., and Fauci, F. 1983, *Nature*, **304**, 417.

Branch, D. 1982, in *Supernovae: A Survey of Current Research*, ed. M. J. Rees and R. J. Stoneham, p. 267 (Dordrecht: Reidel).

Buchler, J.-R., and Mazurek, T. J. 1975, *Mem. Soc. Roy. Sci. Liège*, 6ᵉ ser., **8**, 435.

Buchler, J.-R., Mazurek, T. J., and Truran, J. 1974, *Comm. Ap. Space Sci.*, **6**, 45.

Colgate, S. A., and McKee, C. 1969, *Ap. J.*, **157**, 623.

Courant, R., and Friedrichs, K. O. 1948, *Supersonic Flow and Shock Waves* (New York: Interscience).

Gunn, J. E., and Ostriker, J. P. 1970, *Ap. J.*, **160**, 979.

Hillebrandt, W. 1982, paper read at the XI Texas Symposium on Relativistic Astrophysics, Austin, TX, December 1982; accepted for publication in *Ann. N.Y. Acad. Sci.*

Iben, I., Jr., and Tutokov, A. V. 1983, preprint (IAP 83-16, Illinois astronomy).

Ivanova, L. N., Imshennik, V. S., and Chechetkin, V. M. 1974, *Ap. Space Sci.*, **31**, 497.

Kirshner, R. P., and Oke, J. B. 1975, *Ap. J.*, **200**, 574.

Kudryashov, A. D., Shcherbatyuk, V. A., and Ergma, E. V. 1979, *Ap. Space Sci.*, **66**, 391.

Lyne, A. G. 1982, in *Supernovae: A Survey of Current Research*, ed. M. J. Rees and R. J. Stoneham, p. 405 (Dordrecht: Reidel).

Mazurek, Meier, D. L., and Wheeler, J. C. 1977, *Ap. J.*, **213**, 518.

Mazurek, T. J., and Wheeler, J. C. 1980, *Fundamentals Cosmic Phys.*, **5**, 193.

Meyerott, R. E. 1980, *Ap. J.*, **239**, 257.

Miller, G. E., and Scalo, J. M. 1979, *Ap. J.*, **233**, 767.

Mueller, E., and Arnett, W. D. 1982, *Ap. J. (Letters)*, **261,** L109.

———. 1984, submitted to *Ap. J.*

Mueller, E., and Hillebrandt, W. 1981, *Astr. Ap.*, **103,** 358.

Nomoto, K. 1982*a*, *Ap. J.*, **253,** 798.

———. 1982*b*, *Ap. J.*, **257,** 780.

Ostriker, J. P., Richstone, D. O., and Thuan, T. X. 1974, *Ap. J. (Letters)*, **188,** L87.

Reimers, D., and Koester, D. 1982, *Astr. Ap.*, **116,** 341.

Roache, P. J. 1976, *Computational Fluid Dynamics* (Albuquerque: Hermosa).

Sugimoto, D., and Nomoto, K. 1980, *Space Sci. Rev.*, **25,** 155.

Taylor, J. H., and Manchester, R. N. 1977, *Ap. J.*, **215,** 885.

Trimble, V. 1975, *Rev. Mod. Phys.*, **47,** 877.

Weidemann, V., and Koester, D. 1983, *Astr. Ap.*, **121,** 77.

Wood, P. R., Bessell, M. S., and Fox, M. W. 1983, *Ap. J.*, **272,** 99.

Woosley, S. A., Arnett, W. D., and Clayton, D. D. 1973, *Ap. J. Suppl.*, **26,** 231.

Wu, C.; Leventhal, M.; Sarazin, C. L.; and Gull, T. R. 1983, *Ap. J. (Letters)*, **269,** L5.

15

The Optical Spectrum
of a Carbon-Deflagration Supernova

David Branch

Abstract

Theoretical optical spectra are computed for the Nomoto-Thielemann (NT) hydrodynamical model of a white dwarf which explodes by means of a carbon deflagration after accreting hydrogen at a rate of 4×10^{-8} solar masses per year. The theoretical spectra are compared with spectra of the Type I supernova 1981b in NGC 4536, observed at maximum light and 17 days later. The theoretical and observed spectra are in good agreement provided that the strong radial stratification of composition in the outer layers ($v \gtrsim 8000$ km s^{-1}) of the present NT model is erased by mixing. It is not yet clear whether the discrepancies which remain after the mixing is introduced are due to the bulk composition in the outer layers of the NT model or to the approximations made in the synthetic spectrum calculations.

15.1 Introduction

The ^{56}Ni model for Type I supernovae (Colgate and McKee 1969) accounts both for the characteristic shape of the light curve (see, for example, Woosley 1984 and references therein) and for the optical spectrum observed at "late" times, greater than $\sim$100 days after maximum light, when the ejecta have become optically thin (Axelrod 1980). The observed width of the light-curve peak tells us that the total amount of ejected matter is on the order of one solar mass ($M_\odot$), and the absolute peak brightness requires that the fraction of ejected mass initially in the form of ^{56}Ni is between 0.1 and 1.0 (e.g., Arnett 1982). These constraints, as well as the apparent absence of neutron stars from the remnants of Type I supernovae (SNe I) in our Galaxy (Helfand 1980; Nomoto and Tsuruta 1981) and the rates at which SNe I are found in the various types of external galaxies (Greggio and Renzini 1983; Miller and Chevalier 1983), appear to be consistent with the idea that the immediate progenitors of SNe I are accreting white dwarfs in binary systems. Although the fraction of accreting white dwarfs which manage to evolve to an explosive condition may turn out to be small (Sutherland and Wheeler 1984; Mac-Donald 1984), most current models of exploding white dwarfs do disrupt completely and eject substantial amounts of freshly synthesized nuclear-statistical-equilibrium (NSE) products, including ^{56}Ni. Both the explosion mechanism and the detailed composition of the ejecta depend, however, on the characteristics of the binary system: the mass and composition of the white dwarf, the composition of the accreted matter, and above all the accretion rate. For example, if a carbon-oxygen white dwarf explodes after accreting hydrogen at a rate on the order of 10^{-8} $M_\odot$ yr^{-1}, it does so by means of inward and outward (supersonic) detonation waves which convert practically the entire star to NSE products, while if a C-O white dwarf explodes after accreting hydrogen at 10^{-7} $M_\odot$ yr^{-1}, a single (subsonic) deflagration wave propagates outward, converts roughly the inner half of the star to NSE products, but burns the outer half only partially, to a mixture of intermediate mass elements from carbon to calcium.

It is difficult to use either the observed light curve or the late-time spectrum to distinguish between detonation and deflagration models, because both can eject acceptable amounts of ^{56}Ni, but the "early-time" spectra of the two models, formed in an atmosphere above an optically thick photosphere in the outer layers of the ejecta, should be entirely different. The early spectra of the well-observed Type I supernova 1981b in NGC 4536 have been interpreted as overlapping, P Cygni–type spectral lines of iron, cobalt, and intermediate mass elements from oxygen to calcium (Branch et al. 1982, 1983). Because the spectrum of SN 1981b is typical of a Type I and the detonation models eject hardly any intermediate mass elements, the implication is that all normal SNe I may be produced by deflagrations. Recently Nomoto and Thielemann (Nomoto 1984a, b) have used a nuclear reaction network to calculate the detailed composition ejected by a specific deflagration model—a C-O white dwarf initially of 1 $M_\odot$ which exploded after accreting hydrogen at a rate of 4×10^{-8} $M_\odot$ yr^{-1}. To test this model against observation, I have begun to compute theoretical optical spectra for the model and to compare them with the early spectra of SN 1981b. In its present form the Nomoto-

Thielemann (NT) model has a strongly stratified composition in its outer layers, but Nomoto (1984a) notes that some convective mixing in the partially burned outer layers, where entropy increases outward, is expected. Since a P Cygni line profile depends not only on the number of absorbers above the photosphere but also, through Doppler shifts, on the velocity interval over which the absorbers are distributed, comparison of the synthetic and observed spectra can give an indication of the amount of mixing that is required.

15.2 Assumptions and Approximations

The early-time synthetic spectra shown by Branch et al. (1982, 1983), as well as two of the spectra shown in this paper, were computed not for a detailed hydrodynamical model, but just on the basis of a simple, general picture of the spectrum-forming regions (hereafter the "standard model"). The supernova is assumed to be spherically symmetric and to expand with velocity proportional to radius. A sharp photosphere emits an optical continuous spectrum characteristic of a blackbody at temperature T_{BB}. The formation of spectral lines in the atmosphere above the photosphere is described by the escape-probability treatment of radiative transfer (Sobolev 1960; Castor 1970). The relative line optical depths at the photosphere are determined by an assumed composition and the approximation of local thermodynamic equilibrium (LTE) at temperature T_{ex} and an electron density consistent with unit optical depth to electron scattering at the photosphere. In the escape-probability treatment the line optical depth at any point depends only on the local number of absorbers rather than on an integral over the atmosphere, so the radial dependence of all line optical depths is taken to vary in proportion to an assumed density law, $\rho(v) \sim v^{-7}$. The composition is independent of velocity, and no allowance is made for radial variations in excitation or ionization. T_{BB}, T_{ex}, and the velocity at the photosphere (V) are fitting parameters. In principle, if the composition is fully specified and the electron density is fixed, the absolute scale of the line optical depths is also determined, but in practice, the latter is treated as a free parameter; this corresponds to not fully specifying the composition (the helium abundance, for example) or to relying on the LTE approximation to give only relative but not absolute optical depths of the lines of interest, which are generally of comparable excitation level. Blending of multiple P Cygni lines is treated by an extension of the exact formulation for doublet lines given by Olson (1982).

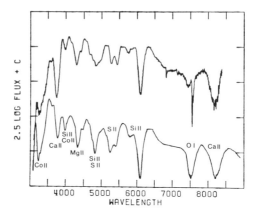

Fig. 15.1. The spectrum of SN 1981b in NGC 4536 observed on 6–9 March 1981 at maximum light (*upper*, Branch et al. 1983) is compared to a synthetic spectrum (*lower*) based on the standard model and the following parameters: $T_{BB} = 20{,}000$ K, $T_{ex} = 11{,}000$ K, $V = 12{,}000$ km s^{-1}, and solar abundances except that Co/Fe = 11. Narrow features near 6800, 7200, and 7600 Å in the observed spectrum are absorptions produced in the Earth's atmosphere. The vertical displacement between the two spectra is arbitrary.

The synthetic spectra for the NT model which are shown in this paper have been computed in the same way, except that (1) the v^{-7} density law has been replaced by that of the model and (2) the radial dependence of the composition in the model has been taken into account in the line optical depths. For each spectrum, the parameters T_{BB}, T_{ex}, and V are held fixed at the values used for the best-fitting spectrum computed with the standard model, while the absolute scale of the optical depths is varied freely. These calculations are preliminary in the sense that in the NT model, unlike the standard model, the composition is fully specified, so the value of T_{ex} and the absolute optical depth scale determine the electron density at the photosphere. For the present calculations the electron density continues to be estimated on the assumption of unit optical depth to electron scattering at the photosphere. Calculations which determine N_e self-consistently will be reported elsewhere (Branch, Nomoto, and Thielemann, in preparation). Because the relative strengths of the important lines in the spectrum depend only weakly on N_e, the present results are expected to be adequate for a first test of the NT model.

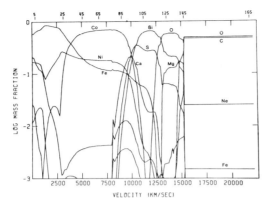

Fig. 15.2. Element abundances in the NT model at 15 days after the explosion are plotted against velocity. Zone numbers in the model are displayed at the top.

15.3 Comparison of Spectra

The maximum-light spectrum of SN 1981b has been compared to spectra computed with the standard model and discussed in detail by Branch et al. (1982) and Branch (1984). A similar comparison is shown here as figure 15.1. The composition is taken to be solar, except that the Co abundance is increased to achieve an absolute Co/Fe ratio of 11, consistent with Co and Fe having originated as ^{56}Ni 15 days earlier, at the time of explosion. The ions responsible for the conspicuous absorption features in the synthetic spectrum are labeled. The synthetic spectrum computed on the basis of this relatively simple model is able to account for all the significant absorption features in the observed spectrum.

Although the density distribution in the NT model is noticeably different from a v^{-7} power law, the effect on the spectrum of inserting the model density distribution, while retaining the assumption of unstratified solar composition, was found to be negligible.

The stratified composition of the NT model is shown in figure 15.2. From the point of view of nucleosynthesis it is customary to display the isotopic composition versus mass fraction (e.g., see Nomoto 1984a), but for spectrum formation it is more informative to plot elemental abundance against velocity. In this model the outermost layers ($\gtrsim$ 15,300 km s^{-1}) consist of the unburned initial composition of the white dwarf (O, C, Ne, and solar mass

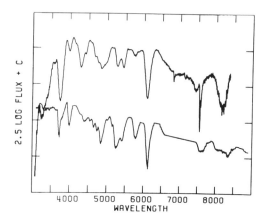

Fig. 15.3. Like figure 15.1, but for the unmixed NT model.

fractions of heavier elements). Products of explosive carbon burning (O, Si, Mg, S) extend from 15,300 down to 13,000 km s^{-1}, and products of oxygen burning (Si, S) dominate from 13,000 to 10,000 km s^{-1}. Shortly after the explosion the NSE products in the innermost region ($\lesssim$ 10,000 km s^{-1}) consist primarily of ^{56}Ni, but figure 15.2 shows the composition 15 days later, at maximum light, by which time beta-decays have transformed most of the nickel to cobalt.

A synthetic spectrum for the NT model at maximum light (figure 15.3) does not match the observed spectrum nearly as well as the spectrum of the standard model. The deterioration of the fit is due mainly to the strong stratification of composition in the NT model rather than to the bulk nonsolar abundance ratios in the outer layers. At this time the photosphere is at 12,000 km s^{-1}, where the model composition is dominated by Si and S. Even though the absolute optical depth scale has been adjusted to make the computed Si and S lines somewhat too strong, the other lines are too weak because the elements are concentrated either beneath (Ca) or too far above (O, Mg) the photosphere.

The composition stratification can be removed by mixing the material in the outer layers of the supernova. Nomoto (1984a) anticipates that the inner, NSE layers will be convectively stable but the outer layers will not be. Therefore, in the absence of a detailed prescription for the mixing, I have completely mixed the outer layers, down to the zone (9600 km s^{-1}) at which

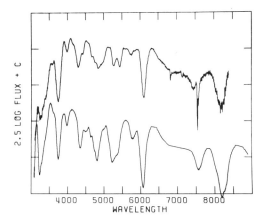

Fig. 15.4. Like figure 15.1, but for the mixed (> 9600 km s^{-1}) NT model.

the mass fraction of NSE products is 0.5. The corresponding spectrum, shown in figure 15.4, is much better than that of the unmixed model (figure 15.3) and matches the spectrum of SN 1981b almost as well as the spectrum of the standard model (figure 15.1). The differences between figure 15.4 and figure 15.1 are due to the nonsolar abundance ratios in the outer layers of this mixed version of the NT model; the relative abundances of Si, S, and Ca are within a factor of 2 of solar, but the O/Si ratio is a factor of 18 lower than in the Sun and the Mg/Si ratio is down by a factor of 8. Owing to the low O/Si ratio, the computed absorption feature near 7500 $\mathring{A}$ is at a longer wavelength than in figure 15.1 and is not really a satisfactory fit to the observations.

The spectrum of SN 1981b observed 17 days after maximum light is compared to a synthetic spectrum for the standard model in figure 15.5. By this time the spectrum has changed significantly from that of maximum light, especially in the blue. Some of the changes can be attributed to a lowering of the level of excitation as the expanding material cools, but to achieve an adequate fit, at least in the LTE approximation, nonsolar composition ratios have had to be introduced. Relative to Si, the Ca, Fe, and Na abundances have been raised by factors of 2.5, 2, and 300, respectively, and the O abundance has been lowered by a factor of 2. The Co/Fe ratio is 4, corresponding to pure initial ^{56}Ni. It is not unreasonable to find the composition near the photosphere to change with time, because the photosphere is receding with respect to the expanding matter, but the very large Na abundance required is

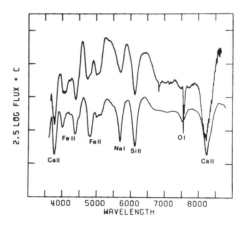

Fig. 15.5. The spectrum of SN 1981b in NGC 4536 observed on 24 March 1981, 17 days after maximum light (*upper*, Branch et al. 1983) is compared to a synthetic spectrum (*lower*) based on the standard model with $T_{BB} = T_{ex} = 9000$ K, $V = 10{,}000$ km s^{-1}, Fe/Si = 2.5, Ca/Si = 2, Na/Si = 300, O/Si = 0.5 (relative to the solar ratios), and Co/Fe = 4 (absolute ratio).

almost certainly due to a breakdown in the LTE assumption for the Na D lines, which, being resonance lines from a neutral atom, are of much lower excitation than the other lines of interest.

At this time the composition of the unmixed NT model is similar to that shown in figure 15.2, except that beta-decays have further altered the composition of the NSE region. The synthetic spectrum for the unmixed NT model, shown in figure 15.6, is not a good fit to the observed spectrum. Again, the stratification is the main problem. The photosphere has receded to 10,000 km s^{-1}, so some of the Ca is now above the photosphere (see figure 15.2), but the Ca does not extend to high enough velocity. In addition, there is not enough Fe above the photosphere and the O is too high in the atmosphere.

If the NT model is mixed, as before, down to 9600 km s^{-1}, the resulting synthetic spectrum contains acceptable Ca lines but the Fe lines remain too weak. To bring more Fe above the photosphere, the model was then mixed deeper, down to 8000 km s^{-1}. In the mixed region the S/Si ratio became solar, Ca/Si was up by a factor of 2, but O and Fe were down by factors of 18 and 8. The synthetic spectrum, shown in figure 15.7, is better, but the Fe lines remain too weak. Obviously, more Fe could be dredged up by mixing to still deeper layers, but a late-time spectrum, 270 days after maximum light,

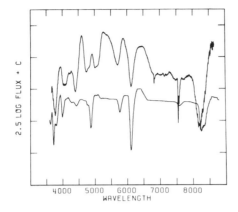

Fig. 15.6. Like figure 15.5, but for the unmixed NT model.

contains a Ca line having a steep red edge corresponding to 8000 km s^{-1}, which implies that the Ca profile extends no deeper into the ejecta (Branch 1984). Mixing Fe up from beneath 8000 km s^{-1} would necessarily mix Ca down into those deep layers, where it is not wanted.

15.4 Discussion

From the point of view of spectroscopy, the carbon-deflagration scenario, unlike the double detonation, is attractive because it produces intermediate mass elements in roughly solar proportions in the outer layers where the early-time optical spectrum is formed. It is clear, however, that the predicted spectrum will be in reasonable agreement with observation only if most or all of the material moving faster than about 8000 km s^{-1} is mixed. Theoretical studies of the mixing mechanism are needed.

As discussed above, the very high sodium abundance needed to match the observed feature is likely to be caused by a breakdown of the LTE approximation. The spectral fits based on the deeply mixed NT model would be improved if the amounts of O, Mg, and Fe relative to Si and S were increased in the outer layers, but it is not yet clear whether this is a deficiency of the supernova model or of the spectrum calculations. The latter are based on the idealization of a sharp photosphere surrounded by a scattering atmosphere (probably not very important in this context), pure resonant scattering in the lines (probably responsible for the fact that the emission components are not matched as well as the absorptions), and the LTE

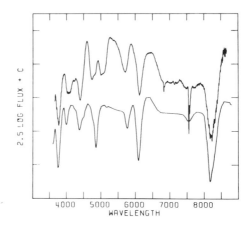

Fig. 15.7. Like figure 15.6, but for the deeply mixed (> 8000 km s^{-1}) NT model.

approximation (possibly important). The characteristics of the NT model could be modified by changing the initial conditions, varying the mixing length and hence the velocity of the deflagration wave, and treating the decay of the deflagration front in the outermost layers in more detail (Nomoto 1984c). A faster deflagration wave would extend the NSE region to higher velocities, and mixing down to 8000 km s^{-1} would then result in a higher Fe/Si ratio in the outer layers. To give a higher O and Mg content, the deflagration front would need to die out more rapidly in the outermost layers.

Considering that the carbon-deflagration scenario is able to account reasonably well for the light curve, the late-time spectra, and the early-time spectra, it seems likely that some variation on this theme is responsible for most, if not all, SNe I. However, several possible problems with the original picture of a C/O white dwarf accreting from a main sequence or giant companion have been pointed out: some sign of hydrogen in the optical spectra of SNe I might be expected if the accreted matter is of ordinary composition, the density in the white dwarf at ignition may be so high as to lead to excessive neutronization of the NSE isotopes, and mass loss during shell flashes may prevent the white dwarfs from evolving to explosive configurations. An interesting way to overcome these obstacles (but presumably to encounter others) is to assume that the donor star, as well as the accretor, is a C/O white dwarf (Iben and Tutukov 1984; Webbink 1984).

Acknowledgments

I am grateful to Ken Nomoto and Friedel Thielemann for providing me with the details of their model, and to Jesse B. Doggett for computational assistance. This work has been supported by NSF grant AST-8218625.

References

Arnett, W. D. 1982, *Ap. J.*, **254**, 1.

Axelrod, T. S. 1980, in *Proc. of the Texas Workshop on Type I Supernovae*, ed. J. C. Wheeler, p. 80 (Austin: University of Texas Press).

Branch, D. 1984, *New York Annals of Science*, **422**, 186.

Branch, D.; Buta, R.; Falk, S. W.; McCall, M. L.; Sutherland, P. G.; Uomoto, A.; Wheeler, J. C.; and Wills, B. J. 1982, *Ap. J. (Letters)*, **252**, L61.

Branch, D.; Lacy, C. H.; McCall, M. L.; Sutherland, P. G.; Uomoto, A.; Wheeler, J. C.; and Wills, B. J. 1983, *Ap. J.*, **270**, 123.

Castor, J. I. 1970, *M.N.R.A.S.*, **149**, 111.

Colgate, S. A., and McKee, C. 1969, *Ap. J.*, **157**, 623.

Greggio, L., and Renzini, A. 1983, *Astr. Ap.*, **118**, 217.

Helfand, D. J. 1980, in *Proc. of the Texas Workshop on Type I Supernovae*, ed. J. C. Wheeler, p. 20 (Austin: University of Texas Press).

Iben, I., Jr., and Tutukov, A. V. 1984, *Ap. J. Suppl.*, **54**, 335.

MacDonald, J. 1984, preprint.

Miller, G. E., and Chevalier, R. A. 1983, *Ap. J.*, **274**, 840.

Nomoto, K. 1984*a*, in *Stellar Nucleosynthesis*, ed. C. Chiosi and A. Renzini (Dordrecht: Reidel), in press.

———. 1984*b*, this volume.

———. 1984*c*, private communication.

Nomoto, K., and Tsuruta, S. 1981, *Ap. J. (Letters)*, **250**, L119.

Olson, G. L. 1982, *Ap. J.*, **255**, 267.

Sobolev, V. V. 1960, *Moving Envelopes of Stars* (Cambridge, MA: Harvard University Press).

Sutherland, P. G., and Wheeler, J. C. 1984, *Ap. J.*, in press.

Webbink, R. F. 1984, *Ap. J.*, **277**, 355.

Woosley, S. E. 1984, this volume.

Nucleosynthesis in Low and Intermediate Mass Stars on the Asymptotic Giant Branch

Icko Iben, Jr.

16.1 Introduction

All single stars initially less massive than about 9 $M_\odot$ and more massive than about 1 $M_\odot$ ultimately evolve onto the asymptotic giant branch (AGB), where they contain a very dense white dwarf–like core composed of carbon and oxygen and have a very extended envelope at the base of which hydrogen and helium burn alternately in shells. Helium burning is initially thermally unstable, and the helium "shell flashing" phase is a dramatic one of short duration: the rate of energy production by helium-burning reactions reaches $(10^7–10^8)$ $L_\odot$, and temperatures at the base of the burning region can exceed 300×10^6 K; neutrons can be released by both the $^{22}\text{Ne}(\alpha,n)^{25}\text{Mg}$ and $^{13}\text{C}(\alpha,n)^{16}\text{O}$ reactions and captured by both the light element progeny of the neutron-producing reactions and by ^{56}Fe and its neutron-capture progeny. Following each helium shell flash (or thermal pulse, as it is often called), products of both α-capture (primarily fresh ^{12}C) and neutron capture (particularly those isotopes produced by capture on ^{56}Fe and its progeny) are brought to the surface where, after a sufficient number of dredge-up episodes has occurred, carbon may exceed oxygen (thus producing a "carbon" star) and distinctive radiative isotopes such as ^{99}Tc and overabundances of other neutron capture products such as Zr (in the form of ZrO) may appear. It is known that AGB stars lose mass at very high rates from their surfaces via a stellar wind and that essentially all of their envelope mass is lost before the central CO core can grow by more than ~ 0.2 $M_\odot$. It is therefore clear that AGB stars can significantly enrich the interstellar medium with carbon and neutron-rich isotopes (in particular those traditionally known as *s*-process isotopes and perhaps also some traditionally referred to as *r*-process isotopes).

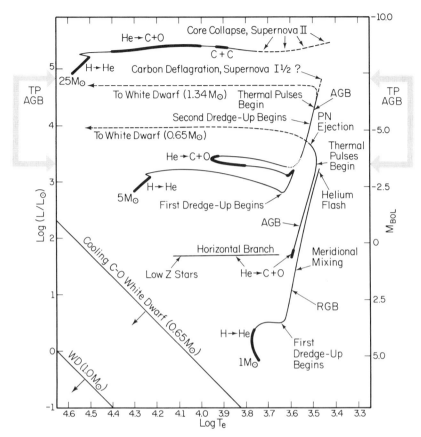

Fig. 16.1. Representative evolutionary tracks for low, intermediate, and high mass stars. Heavily weighted portions indicate where nuclear burning in the stellar core takes place on a long time scale. Note that, for the 5 $M_\odot$ star, core He burning occurs in two distinct regions: one essentially on the red giant branch (not labeled) and one considerably to the blue of this branch (labeled). The 5 $M_\odot$ track actually exhibits considerable "looping" following core He burning. This looping, which occurs on a very short time scale, has been suppressed and replaced by a dotted track. The dashed portions that comprise the ends of the three tracks are schematic only. For the 1 and 5 $M_\odot$ stars, exactly where each track bends to the left and proceeds blueward to the region of white dwarfs depends on the average rate of mass loss the star experiences and on when, if ever, a planetary nebula is ejected. Where the 25 $M_\odot$ star spends its time while developing a core of Fe-peak elements is also a function of the mass it has lost. Tracks for possible white dwarf progeny of AGB stars are shown schematically (they are actually at smaller luminosities than shown). Thermally pulsing AGB phases are confined to the narrow interval $-3.5 \gtrsim M_{bol} \gtrsim -7.3$. (I. Iben, Jr., and A. Renzini, 1983, *Ann. Rev. Astr. Ap.,* **21,** 271.)

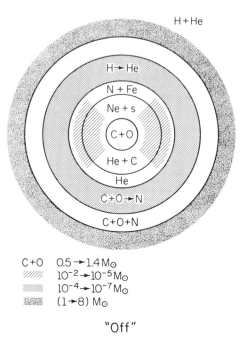

H+He

H→He

N + Fe

Ne + s

C + O

He + C

He

C+O→N

C+O+N

C+O	$0.5 \rightarrow 1.4 \, M_\odot$
/////	$10^{-2} \rightarrow 10^{-5} \, M_\odot$
	$10^{-4} \rightarrow 10^{-7} \, M_\odot$
	$(1 \rightarrow 8) \, M_\odot$

"Off"

Fig. 16.2. Zones of differing composition and nucleosynthetic activity during the "off" phase. Helium burning is essentially extinct, and hydrogen-burning reactions provide nearly all of the energy appearing at the surface.

Given the importance of AGB stars for producing carbon and neutron-rich isotopes, it is astonishing that they not only occupy an incredibly small region in the H-R diagram ($3.5 \lesssim \log L < 4.8$, $-7.3 < M_{\mathrm{bol}} \lesssim -4$, $3.4 \lesssim \log T_e \lesssim 3.6$) but, compared with their lifetime on the main sequence as core hydrogen burners, exist as element manufacturers for an incredibly short time ($\sim 2 \times 10^6$ yr for low mass AGB stars and less than $\sim 10^5$ yr for intermediate mass AGB stars with $M_{\mathrm{bol}} < -6$). Their location in the H-R diagram is shown in figure 16.1.

16.2 Thermal Pulses, an Overview

From the point of view of nucleosynthesis, the life of an AGB star may be divided into five phases: (1) an "off" phase, when only hydrogen is burning; (2) an "on" phase, when helium burning reaches its peak and leads to the formation of a convective shell; (3) a "power-down" phase, when the overpressure produced by the sudden injection of nuclear energy causes matter in both the helium-burning and hydrogen-burning regions to expand outward and cool—the rate of helium burning drops precipitously and hydrogen burning effectively ceases; (4) a "dredge-up" phase, when products of nucleosynthesis in the helium-burning convective shell are carried to the surface; and (5) a quiescent helium-burning phase, when the majority of the helium that has been produced by hydrogen burning during the off phase is converted completely into carbon and oxygen.

The distribution of elements and of dominant nuclear reaction processes during phases 1, 2, and 4 are shown in the circular schematics of figures 16.2, 16.3, and 16.6. The rectangular schematics of figure 16.5 show element distributions and reaction processes during these three phases and during phase 3 as well. A schematic for the quiescent helium-burning phase would differ from that for the dredge-up phase only in that the base of the convective envelope moves slowly outward in mass and there is no net change in element and isotopic abundances in the envelope.

In figure 16.2, the base of the extensive convective envelope (whose inner radius is of dimensions of the order of 1–10 $R_\odot$ and whose outer radius can reach ~ 500–1000 $R_\odot$) is depicted by the outer "iron-filing" ring. Through the next inward unshaded region, energy flows outward by radiative diffusion from the region of dominant energy production shown by regularly spaced dots. In this dotted region (typically at a radius of from 0.01 to 0.1 $R_\odot$), hydrogen is converted completely into helium and both ^{12}C and ^{16}O are converted almost completely into ^{14}N. Immediately below the actively burning region is a region containing the ashes of pure hydrogen burning (^{4}He, ^{14}N) and isotopes such as ^{56}Fe that survive quiescent hydrogen burning. Next is a region containing products of incomplete helium burning that are left over from a previous helium shell flash. This region contains ^{22}Ne (a result of two α-capture reactions and two β-decays beginning with ^{14}N) and products of neutron capture nucleosynthesis denoted by s (for s-process). Finally, there is the central core of inert carbon and oxygen which is kept at temperatures of $\sim (100$–$400) \times 10^{6}$ K by a balance between neutrino losses and the release of gravitational potential energy due to compression, as the ashes of nuclear burning in the outer shells increase the mass of the white dwarf–like core.

Compressional heating also operates in the cross-hatched helium-rich regions, and, when the mass of material in these regions (recurrently) exceeds a critical value, helium burning "runs away," with consequences illustrated in

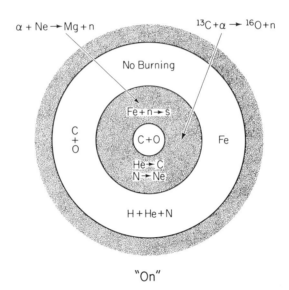

$\alpha + Ne \rightarrow Mg + n$ $^{13}C + \alpha \rightarrow {}^{16}O + n$

Fig. 16.3. Zones of differing composition and nucleosynthetic activity during the "on" phase of a thermal pulse cycle. Helium-burning reactions are intense $(L_{He} \rightarrow 10^7\text{–}10^8\ L_\odot)$ and force convection to extend from the base of the burning zone nearly to the hydrogen-helium interface.

figure 16.3. The rate of helium burning reaches a peak as large as $(10^7\text{–}10^8)$ $L_\odot$, depending on core mass and on the number of previous pulses which have occurred. The flux of energy is so high that a convective shell is built up (the "iron-filing" region surrounding the CO core). The outer edge of this shell reaches almost to the inner edge of the hydrogen-rich region but is prevented from extending into this region by an "entropy" barrier, as shown in figure 16.4. Thus, hydrogen is not injected into the helium-burning shell where it can interact with the ^{12}C there to initiate the neutron-producing reactions $^{12}C(p,\gamma)^{13}N(\beta^+\nu)^{13}C(\alpha,n)^{16}O$.

Temperatures at the base of the convective shell reach as high as $(300\text{–}350) \times 10^6$, and in all cases the ^{14}N which enters the shell is quickly converted into ^{22}Ne by the reactions $^{14}N(\alpha,\gamma)^{18}F(\beta^+\nu)^{18}O(\alpha,\gamma)^{22}Ne$. However, only in stars with core mass as large as $\sim 0.9\ M_\odot$ do temperatures remain high enough, long enough, for the $^{22}Ne(\alpha,n)^{25}Mg$ reaction to produce a significant flux of neutrons. This statement assumes, of course, that current estimates of the rate of the $^{22}Ne(\alpha,n)^{25}Mg$ reaction are correct. As will be described in § 16.4, it is possible that, in stars of small core mass, ^{13}C built up

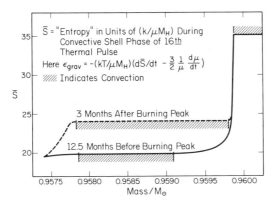

Fig. 16.4. The entropy parameter $\bar{S}$ versus mass during the growth phase of a convective shell during the 16th pulse of a model of total mass $M = 7\ M_\odot$ and core mass $M_{core} \sim 0.96\ M_\odot$. The helium-burning rate peaks near the base of the convective shell, and the most rapid rise in entropy occurs through the hydrogen-burning shell. (I. Iben, Jr., 1976, *Ap. J.*, **208,** 165.)

immediately following any given shell flash may be swept into the helium-burning convective shell during the following shell flash and the $^{13}C(\alpha,n)^{16}O$ may then provide an abundant source of neutrons.

The nuclear energy released suddenly in the form of γ-rays during pulse peak is converted immediately into thermal motions of nuclei and electrons, and the overpressure thereby produced forces matter in the burning region to expand and cool. Expansion and cooling propagates out to where there is hydrogen, and the conversion of hydrogen effectively ceases. As shown in the third panel from the left in figure 16.5, the reactions that take place in the convective shell during pulse peak continue during this "power-down" phase, but they proceed at significant rates only in the inner half of the helium-rich zone that was earlier contained in the convective shell and are effectively extinguished in the outer half of this zone.

Toward the end of the power-down phase, the energy that is being released by helium-burning reactions finally breaks out of the production zone via radiative diffusion. The matter in the innermost portion of the helium-burning region now contracts as the star adjusts to a new quasi-static equilibrium between nuclear energy production and surface energy loss. For a time, the helium-burning luminosity increases (this may be thought of as an overshoot of equilibrium), but, after reaching a secondary maximum that is

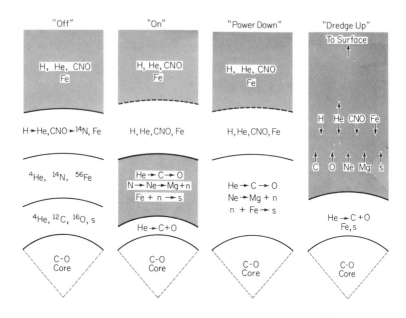

Fig. 16.5. Schematic representation of the composition distribution, the dominant nuclear reactions, and the convective regions (*shaded*) during (1) the "off" phase between thermal pulses, (2) the "on" phase of a pulse, and the (3) early and (4) late (dredge-up) phases of a pulse power-down. The region between the C-O core and the hydrogen-rich convective envelope is highly exaggerated and distorted in size relative to the sizes of the core and envelope. For example, when $M_{CO} \sim 1\ M_\odot$, the hydrogen-burning shell (H → He, CNO → ^{14}N) is of mass $\sim 10^{-5}\ M_\odot$; the mass of the region containing unprocessed matter and fresh products of hydrogen burning (^{4}He, ^{14}N, ^{56}Fe) grows to roughly $10^{-3}\ M_\odot$, and the mass of the region containing products of processing in a convective shell during the preceding "on" phase (^{4}He, ^{12}C, ^{16}O, s) is also of the order of $10^{-3}\ M_\odot$. During the "off" phase, helium-burning reactions are effectively extinguished (region ^{4}He, ^{12}C, ^{16}O). During the "on" phase, hydrogen burning is extinguished and the conversions 3 ^{4}He → ^{12}C, ^{12}C + ^{4}He → ^{16}O, ^{14}N + 3 ^{4}He → ^{25}Mg + n, ^{56}Fe + n → ^{57}Fe, etc. (= s-process isotopes), take place in a convective shell that contains both matter laid down by the hydrogen-burning shell between pulses and matter processed in the convective shell during the preceding pulse. During the early portion of the power-down phase, convection in a shell ceases, but helium-burning and neutron capture reactions continue to take place. Then, as these reactions decrease in frequency, convection extends into the region that has been freshly processed during the pulse "on" phase and mixes this matter uniformly throughout the envelope. Thus, fresh carbon and s-process elements are brought to the surface. This is the "dredge-up" phase. (I. Iben, Jr., 1981, in *Physical Processes in Red Giants*, ed. I. Iben, Jr., and A. Renzini, p. 1 [Dordrecht: Reidel].)

perhaps a factor of 2 larger than the hydrogen-burning luminosity during the "off" phase, it declines steadily.

During the increase in helium-burning luminosity toward the secondary maximum, matter in the outer portion of the carbon-rich region formed during pulse peak and all matter beyond continues to expand outward and cool. As the temperature and density at the base of the convective envelope decrease, the opacity there increases. In response to this decrease and to the increasing flux of energy through this region, matter successively farther inward becomes unstable toward convection. Hence, the base of the convective envelope moves inward in mass, ultimately reaching the interface between hydrogen-rich and hydrogen-exhausted matter. Convection then carries products of complete hydrogen burning to the surface. Soon, the base of the convective envelope reaches the outer edge of the region earlier found in a convective shell and products of incomplete helium burning and of neutron capture are swept to the surface. This is the dredge-up phase depicted in the rightmost panel of figure 16.5 and in figure 16.6.

Ultimately, when the rate of energy production by helium burning reaches a (secondary) maximum and thereafter declines, dredge up ceases and quiescent helium burning continues for approximately 10% of the interpulse lifetime (the energy released by one gram of helium is about 10% of that released by one gram of hydrogen, and the luminosity of the star is, to first approximation, constant). During this quiescent helium-burning phase, the interface between hydrogen-rich and hydrogen-free matter moves slowly inward in radius (the interface mass is, of course, fixed), rising in temperature and density, and the rate of hydrogen burning slowly increases. Eventually, hydrogen burning takes over completely from helium burning in supplying the flux of energy at the surface of the star and the cycle begins again.

16.3 Neutron Capture in AGB Stars of Large Core Mass: The ^{22}Ne Source

In thermally pulsing stars of large core mass, there are three major properties of the overall neutron capture process that conspire to produce neutron capture isotopes in the solar system distribution. The first of these properties is the overlap of successive convective shells. That is, a fraction of the matter appearing in the convective shell during any given pulse peak has appeared in the convective shell during the previous pulse peak. This is

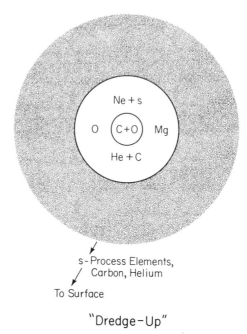

s-Process Elements,
Carbon, Helium
To Surface

"Dredge-Up"

Fig. 16.6. The dredge-up phase. This occurs when the energy injected during pulse peak has pushed matter at the hydrogen-helium interface out to very low densities and temperatures and when the flux through this interface increases due to a secondary increase in the helium-burning luminosity.

illustrated in figure 16.7. If we call the "overlap" fraction r, then the fraction of matter appearing in the Nth convective shell that has been within M previous convective shells is $r^M = e^{M \ln r} = e^{-M|\ln r|}$, where M can go from 1 to $N - 1$. If the "neutron exposure" per flashing episode ($=$ number of neutrons emitted per shell event, per iron seed) is constant (or varies slowly) from one convective shell episode to the next, then it is clear that an exponential distribution of exposures results. Thus, one essential requirement for producing s-process isotopes in the solar system distribution—an exponential distribution of neutron exposures—is met in a very natural and straightforward way in thermally pulsing AGB stars.

The second essential property is a sufficient flux of neutrons per ^{56}Fe seed nucleus. This too follows in a very natural and straightforward way in thermally pulsing AGB stars of large core mass, wherein the ^{22}Ne$(\alpha,n)^{25}$Mg reaction goes essentially to completion, so that the number of neutrons released

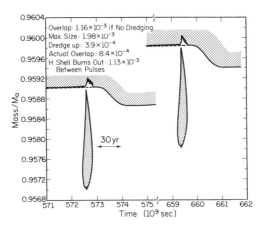

Fig. 16.7. Convective regions during the 15th and 16th pulses in a 7 $M_\odot$ model of core mass $M_{core} \sim 0.96$ $M_\odot$. The shaded regions indicate where matter is convective. The dashed lines indicate the location of the hydrogen-helium interface before the dredge-up phase. (I. Iben, Jr., 1976, *Ap. J.*, **208**, 165.)

per gram of matter in a convective shell is effectively equal to the number of CNO nuclei per gram with which the main sequence progenitor of the AGB star began. Since there are roughly 40 CNO nuclei per ^{56}Fe nucleus in unprocessed matter in typical Population I stars, the number of neutrons per iron seed is actually far too large for the production of s-process isotopes in the solar system distribution, *if all of the neutrons were to be captured by* ^{56}Fe *and its progeny.*

What saves the day and, in fact, conspires to produce "just the right" number of neutrons per iron seed is the fact that ^{22}Ne, ^{25}Mg, and the neutron capture progeny of these two isotopes capture most of the neutrons released, leaving just the right number to be captured by ^{56}Fe and its progeny. This is the third essential property.

It can be shown that, after a sufficient number of pulses, the final isotopic distribution in a convective shell reaches a "steady state" (which of course varies slowly as the core mass increases) and that any two adjacent isotopes in the neutron capture chain are related to one another by

$$\sigma_i n_i = (1 + \Lambda/\sigma_i)^{-1} \sigma_{i-1} n_{i-1} , \qquad (1)$$

where σ_i is the average neutron capture cross section of the ith isotope, n_i is the steady state abundance of the ith isotope, and Λ is a "universal" parameter which may be approximated by

$$\Lambda \sim \sigma_l + [(n_s/n_0)\sigma_s + (n_{l'}/n_0)\sigma_{l'} + (n_{22}/n_0)\sigma_{22}]$$
$$\times [1 + (1 - r)(1 + r)^{-1}(2/b - 1)] \quad . \tag{2}$$

In this last expression, the σ's are the average neutron capture cross sections and the n's are the number densities of (1) ^{25}Mg and its neutron capture progeny (l); (2) the light elements brought into the convective shell after passing through the hydrogen-burning shell between pulses (l'); (3) ^{22}Ne (22); and ^{56}Fe and its neutron capture progeny (s). The degree of overlap, r, has already been defined, b (for burned) is the fraction of ^{22}Ne that experiences α-capture in any given pulse, and n_0 is the number density of CNO elements at the stellar surface (= the number density of ^{22}Ne formed early on in a thermal pulse).

Although the parameter Λ which defines the nature of the neutron capture distribution appears to be predominantly a function of the nuclear properties of the relevant isotopes, it is the recurrence property of the thermal pulses that has made this functional form possible. In typical AGB stars of large core mass, the quantity in the second square bracket of equation (2) is near unity. Further $(n_s/n_0)\sigma_s \sim (1/40)$ 15 mb ~ 0.4 mb; $(n_{l'}/n_0)\sigma_{l'} \sim (1/2)$ 3 mb; $(n_{22}/n_0)\sigma_{22} \sim \sigma_{22}/2$; and $\sigma_l \sim (3-5)$ mb. Hence

$$\Lambda \sim \sigma_l + (1/2)(\sigma_{l'} + \sigma_{22}) \sim \sigma_{\text{lights}} \quad , \tag{2'}$$

and we have the remarkable result that, in first approximation, the parameter characterizing the distribution of s-process isotopes which is built up by thermally pulsing stars of large core mass is the average cross section for neutron capture by ^{22}Ne, ^{25}Mg, and the neutron capture progeny of these isotopes. It is nearly independent of the abundances of ^{56}Fe and of its neutron capture progeny! The essential reason for this result is that (for nearly complete ^{22}Ne burning) the number of light element "filters" is equal to the number of neutrons released. The number of neutrons captured per ^{56}Fe seed to form s-process isotopes is nearly constant and independent of n_s: $\sigma_s/\sigma_{\text{lights}} \sim 15/(3-5) \sim 3-5$.

In concluding this section, it must be cautioned that it is at present highly uncertain as to whether real AGB stars of large core mass survive long enough to produce s-process isotopes in sufficient abundance to contribute significantly to the enrichment of the interstellar medium in these isotopes. In the Magellanic Clouds, the observational counterparts of AGB models of large core mass appear not to live long enough to develop a surface C/O ratio larger than unity and, in fact, probably eject a planetary nebula and become white dwarfs after living only about 10^5 yr as AGB stars. This lifetime would appear to be perhaps a factor of 10 less than is necessary to produce a Galaxy-wide abundance of s-process isotopes that is in the solar system ratio with respect to, say, iron. One may infer that either (1) our own solar system distribution of isotopes is the consequence of the contribution to the matter in the solar nebula by a single nearby AGB star; (2) the $^{22}Ne(\alpha,n)^{25}Mg$ cross section has been underestimated by one or two orders of magnitude so that AGB stars of low core mass (models of which currently burn only about 1% of their ^{22}Ne, but for which many observational counterparts exist) can contribute to the galactic nucleosynthesis of s-process isotopes in the solar system distribution; or (3) the maximum temperature reached at the base of the convective shell in current models of small core mass is 10–15% smaller than that in real counterparts.

16.4 Dredge Up and Neutron Capture in AGB Stars of Small Core Mass: The ^{13}C Source

Just as one may characterize the situation with regard to stars of large core mass as one of a beautiful theory in search of observational counterparts, so one may characterize the situation with regard to stars of small core mass as one of real stars in search of a theoretical explanation. Carbon stars are thought to be low mass AGB stars which have dredged up carbon that has been manufactured by the triple-α process in a convective helium-burning shell, and yet, until quite recently, theoretical models did not experience dredge up for small enough core masses to account quantitatively for the observations. In the Magellanic Clouds, for example, the bulk of the carbon stars have magnitudes lying in the range $-4 > M_{bol} > -6$, where the upper limit (maximum $|M_{bol}|$) corresponds to a core mass of about 0.8 $M_\odot$ and the lower limit corresponds to a core mass of about 0.6 $M_\odot$. For the entire period from the discovery of thermal pulses in 1965 to the year 1982, no models of appropriately low core mass and total mass exhibited the property of dredge up of products of incomplete helium burning.

In 1982 it was shown that, in stars of low metallicity (appropriate for stars in the Magellanic Clouds), two effects are important for achieving dredge up in model counterparts of Cloud stars. First, since pulse amplitude grows

with each pulse and since the probability and extent of dredge up is proportional to some power of pulse amplitude, it is necessary to follow explicitly the
evolution of the model star through all phases prior to achieving any given
core mass. That is, an AGB star possesses "memory" and one must therefore
make sure that evolutionary calculations begin with the proper initial conditions.

Second, the extent of expansion and cooling of matter at the edge of the
carbon-rich region and therefore the extent to which the base of the convective envelope penetrates this region is influenced by whether or not bound-free
transitions from K-shell electrons in ^{12}C are of importance. It turns out that,
when one follows carefully the evolution of a model star of low initial total
mass ($M \lesssim 2\ M_\odot$) from the onset of the thermally pulsing phase (when core
mass $M_{core} \approx 0.53\ M_\odot$), the amplitude of a given pulse becomes large enough
when $M_{core} \sim 0.6\ M_\odot$ that the outer edge of the ^{12}C-rich region built up during the convective shell phase is carried out to low enough temperatures (~ 3
$\times 10^6$ K) that carbon partially recombines and the opacity becomes locally
nearly proportional to the ^{12}C abundance. The net result is an additional
opacity "blocking" that forces the carbon-rich material and matter beyond to
even lower densities and temperatures than it would otherwise (neglecting the
bound-free transitions from ^{12}C) have experienced. The base of the convective envelope now moves inward (in mass) to dredge freshly made carbon up
to the surface. This process is illustrated in figure 16.8.

Also shown in figure 16.8 is an additional consequence of tapping an opacity source which is proportional to an abundant element (namely ^{12}C). Just
prior to the dredge-up phase, the high opacity due to recombining carbon
leads to convective motions which mix carbon outward, thereby reducing the
opacity. The net result is the formation of a classical semiconvective zone
wherein ^{12}C is distributed in such a way that the radiative and adiabatic gradients are equal to each other (see figure 16.9). Of course, with ^{12}C
"diffusing" outward, hydrogen also "diffuses" inward and this latter diffusion
continues during the dredge-up phase. When dredge up is completed, there
exists a zone containing ^{12}C at a relatively large abundance and ^{1}H at a relatively small abundance.

It is this "CH" zone that is ultimately responsible for the activation of
the ^{13}C neutron source. After the dredge-up phase is completed and during
the quiescent helium-burning phase, this zone (along with all other matter in

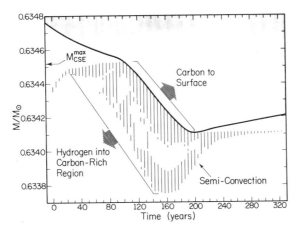

Fig. 16.8. Convective and semiconvective zones in a model AGB star of small total mass (0.7 $M_\odot$) and of small core mass following the 14th pulse. Metallicity Z = 0.01. Full-scale convection occurs in the dotted region which extends outward to the photosphere. The formal convection zones in the semiconvective region are indicated by vertical lines for every fifth model. (I. Iben, Jr., and A. Renzini, 1982, *Ap. J. [Letters]*, **263**, L188.)

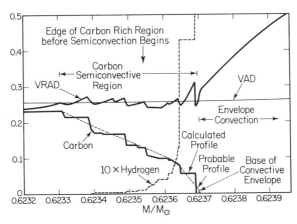

Fig. 16.9. Composition and temperature gradient profiles just before the outer edge of the carbon semiconvective region and the inner edge of the envelope convective zone touch. Same model as described in figure 16.10. Carbon and hydrogen abundances by mass are shown, as are VRAD = $(d \log T/d \log P)_{rad}$ and VAD = $(d \log T/d \log P)_{ad}$. (I. Iben, Jr., and A. Renzini, 1982, *Ap. J. [Letters]*, **259**, L79.)

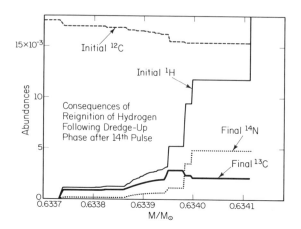

Fig. 16.10. Composition profiles between the base of the semiconvective zone and the base of the convective envelope at the maximum inward extent of each of these regions. Number abundances per gram are obtained by multiplying by Avogadro's number. The abundances of ^{12}C and ^{1}H are shown prior to hydrogen burning, whereas abundances of ^{13}C and ^{14}N are a consequence of hydrogen burning. (I. Iben, Jr., and A. Renzini, 1982, *Ap. J. [Letters]*, **263**, L188.)

the envelope) slowly contracts and heats until eventually proton capture reactions begin. Over a substantial portion of the CH zone, there are only enough protons to convert ^{12}C into ^{13}N and nuclear reactions end with $^{13}N(\beta^{+}\nu)^{13}C$. Over the outer portion of the zone, an additional proton capture leads to the production of ^{14}N at an abundance greater than ^{13}C. Quantitative results for a particular case are shown in figure 16.10.

When the next pulse occurs, the ^{13}C produced in the CH zone is engulfed by the convective shell as it advances outward (in mass) toward the hydrogen-helium interface (which it is prevented from entering by an entropy barrier). The rate at which neutrons are released in the convective shell by the $^{13}C(\alpha,n)^{16}O$ reaction is determined, not by the intrinsic rate at which the α-capture reaction occurs, but by the rate at which fresh ^{13}C enters the convective shell. In the particular example illustrated in figure 16.10, roughly 30 neutrons are released per ^{56}Fe seed.

The physics of mixing during the combined dredge-up and semiconvective stage has just begun to be explored, and no explicit neutron capture calculations relevant to the environment provided by thermally pulsing stars of small core mass have been made. It is therefore premature to speculate extensively on the possible outcome of quantitative studies. It does appear at this point, however, that the distribution of neutron-rich isotopes formed in any given model will vary widely from one model to the next, depending sensitively on core mass, total stellar mass, metallicity, and so on. AGB stars of large core mass remain the most likely source of s-process isotopes in the solar system distribution, despite their apparent rarity of occurrence in nature.

Acknowledgment

This research has been supported in part by National Science Foundation grant AST 81-15325.

Bibliography

This bibliography is far from complete and does not do justice to the many dedicated scientists who have contributed to this fascinating field. It is designed only to provide a starting point for the interested reader.

A.   Stellar Structure: Instability and Mixing in AGB Stars

1.   Schwarzschild, M., and Härm, R. 1965, *Ap. J.,* **142,** 855; 1967, *Ap. J.,* **150,** 961. Demonstrates the occurrence of a thermal instability during the AGB phase in stars of small core mass.

2.   Weigert, A. 1966, *Zs. f. Ap.,* **64,** 395. Demonstrates the occurrence of the same instability in AGB stars of large core mass.

3.   Gingold, R. A. 1974, *Ap. J.,* **193,** 177. Demonstrates that products of complete hydrogen burning are dredged up in models of low core mass.

4.   Iben, I., Jr. 1975, *Ap. J.,* **196,** 525; 1966, *Ap. J.,* **208,** 165. Demonstrates that the ^{22}Ne neutron source is activated and that products of partial helium burning and of s-processing are dredged up in models of large core mass.

5.   Iben, I., Jr. 1976, *Ap. J.,* **208,** 165; 1977, *Ap. J.,* **217,** 788. Demonstrates that hydrogen is prevented from entering the helium-burning convective shell by an entropy barrier.

6.   Wood, P. R. 1981, in *Physical Processes in Red Giants,* ed. I. Iben, Jr., and A. Renzini, p. 135 (Dordrecht: Reidel). Demonstrates the sensitivity of dredge-up properties to total stellar mass, opacity source (e.g., metallicity), and treatment of envelope convection (mixing length/scale height).

7.   Becker, S. A. 1981. In *Physical Processes in Red Giants,* ed. I. Iben, Jr., and A. Renzini, p. 141 (Dordrecht: Reidel). Compares and catalogs

results of thermal pulse calculations by many authors and reports that the ^{22}Ne source is activated (but only at the 1% level) in AGB models of small core mass.

8. Iben, I., Jr., and Renzini, A. 1982, *Ap. J. (Letters)*, **259,** L79; 1982, *Ap. J. (Letters)*, **263,** L188. Demonstrates the role of partial carbon recombination in producing the dredge-up products of helium burning and of neutron capture in stars of small core mass and small metallicity.

9. Iben, I., Jr., and Renzini, A. 1982, *Ap. J. (Letters)*, **263,** L188. Demonstrates that, following the semiconvective phase that accompanies dredge up after any given pulse, hydrogen burning produces a residue of ^{13}C which is incorporated in the convective shell of the next pulse. A substantial number of neutrons are then released by the ^{13}C$(\alpha,n)^{16}$O reaction.

10. Iben, I., Jr. 1983, *Ap. J. (Letters)*, **275,** L65. Demonstrates the dredge up of helium-burning and neutron capture products in models of small core mass when metallicity is high, demonstrates that dredge up has an "on-off" character, and reports that the rate at which neutrons are released by the ^{13}C$(\alpha,n)^{16}$O reaction is governed by the rate at which ^{13}C is incorporated into the convective shell rather than by the intrinsic rate of the α-capture reaction.

B. Neutron Capture Nucleosynthesis in the Environment of a Thermally Pulsing AGB Star

1. Cameron, A. G. W. 1955, *Ap. J.*, **121,** 144. Identifies ^{13}C$(\alpha,n)^{16}$O as a potentially important neutron source.

2. Burbidge, E. M.; Burbidge, G. R.; Fowler, W. A.; and Hoyle, F. 1957, *Rev. Mod. Phys.*, **29,** 547. Identifies the distinctive character of the *s*-process and of the *r*-process.

3. Cameron, A. G. W. 1960, *A.J.*, **65,** 485. Identifies ^{22}Ne$(\alpha,n)^{25}$Mg as a potentially important neutron source.

4. Clayton, D. D.; Fowler, W. A.; Hull, T. C.; and Zimmerman, B. A. 1961, *Ann. Phys.*, **12,** 121. Shows the essential importance of an exponential distribution of neutron exposures in producing *s*-process isotopes in the solar system distribution.

5. Ulrich, R. K. 1973, in *Explosive Nucleosynthesis,* ed. D. N. Schramm and W. D. Arnett, p. 139 (Austin: University of Texas). Points out that the

overlap of matter appearing in successive convective shells in thermally pulsing stars leads to an exponential distribution of exposures, identifies the appropriate number of neutrons to be captured by ^{56}Fe and its progeny per pulse in order to achieve a solar system distribution of s-process isotopes, and develops a formalism containing the parameter Λ.

6. Iben, I., Jr. 1975, *Ap. J.*, **196**, 549. Demonstrates that ^{22}Ne and ^{25}Mg and the neutron capture progeny of these two isotopes act as filters to permit just the right number of neutrons to be captured by ^{56}Fe and its progeny to produce the solar system s-process distribution and further develops the formalism by relating Λ to nuclear and stellar characteristics.

7. Truran, J. W., and Iben, I., Jr. 1977, *Ap. J.*, **216**, 797. Demonstrates that a steady state distribution as postulated in items 5 and 6 of this section of the bibliography can be built up after a finite number of pulses and obtains a general expression (see this text) for Λ as a function of nuclear and stellar characteristics.

8. Despain, K. H. 1980, *Ap. J. (Letters)*, **236**, L165. Shows that the average neutron flux during a thermal pulse is too large (by several orders of magnitude) to produce all s-process isotopes in the solar system distribution.

9. Ward, R. A. 1977, *Ap. J.*, **216**, 540. Shows that isomeric levels in several critical nuclei may not thermalize on time scales short compared to the duration of a pulse.

10. Cosner, K., Iben, I., Jr., and Truran, J. W. 1980, *Ap. J. (Letters)*, **238**, L91. Shows that the time dependence of the neutron flux (decaying exponentially with time) plays a crucial role in determining final abundances and that the net result is still a solar system–like s-process distribution. Several apparent glaring departures from the solar system distribution are removed if the suggestions of Ward (item 9) are applicable.

11. Almeida, J., and Käppeler, F. 1983, *Ap. J.*, **265**, 417. Demonstrates that a solar system distribution of s-process isotopes is formed even with a cross section for the ^{22}Ne$(n,\gamma)^{23}$Ne reaction which is an order of magnitude larger than that derived by a v^{-1} extrapolation of the thermal cross section.

C. Observational Tests

Important work by many authors including Aaronson, Bessel, the Blancos, Cohen, Elias, Fox, Frogel, Little, Little-Marenin, Lloyd-Evans, McCarthy, Mould, Persson, Richer, Westerlund, and Wood is surveyed in:

1. Several articles in 1984, *Proc. IAU Symposium 105: Observational Tests of Stellar Evolution Theory*, ed. A. Maeder and A. Renzini (Dordrecht: Reidel).

2. Articles by V. M. Blanco, M. F. McCarthy, J. A. Frogel, J. G. Cohen, S. E. Persson, J. H. Elias, H. Richer, and J. M. Scalo in 1981, *Physical Processes in Red Giants,* ed. I. Iben, Jr., and A. Renzini (Dordrecht: Reidel).

3. A review by I. Iben, Jr., and A. Renzini in 1981, *Ann. Rev. Astr. Ap.,* **21,** 271.

   Particularly crucial studies are:

4. Blanco, V. M., McCarthy, M. F., and Blanco, B. M. 1980, *Ap. J.,* **242,** 938. This shows that the carbon star distribution in the Magellanic Clouds extends to smaller magnitudes (by $|\Delta M_{bol}| \sim 1$) than previously predicted by theory and cuts off at smaller magnitudes (also by $|\Delta M_{bol}| \sim 1$) than previously predicted.

5. Mould, J. R., and Aaronson, M. 1982, *Ap. J.,* **263,** 629; Lloyd-Evans, T. 1983, *M.N.R.A.S.,* **204,** 985; Frogel, J. A., and Blanco, V. M. 1984, in *Observational Tests of Stellar Evolution Theory,* ed. A. Maeder and A. Renzini (Dordrecht: Reidel), in press. These show that, in old clusters ($\sim 10^{10}$ yr) in the Magellanic Clouds, AGB stars die before core mass becomes large enough for dredge up to occur. In clusters of intermediate age (3×10^8 yr $\rightarrow 3 \times 10^9$ yr), the transition luminosity between M stars (O > C at the surface) and C stars (C > O) increases with decreasing age, as predicted by the theory. In the youngest clusters ($\lesssim 10^8$ yr), however, not only do C stars not occur for $M_{bol} < -6$, but neither do M stars!

6. Wood, P. R., Bessel, M. S., and Fox, M. W. 1983, *Ap. J.,* **272,** 99. This shows that, despite the apparent lack of AGB stars with $M_{bol} < -6$ in clusters in the Magellanic Clouds, there is a class of long period variables with $-7 \lesssim M_{bol} \lesssim -6$ that is distinguished by exhibiting surface abundances of ZrO larger than normal. The total number of such objects

may be used to estimate a lifetime of $\sim 10^5$ yr for an AGB star with M_{bol} < -6.

7.   Smith, V. V., and Wallerstein, G. 1983, *Ap. J.*, **273,** 742. This demonstrates that more extensive studies of the distribution of neutron-rich elements such as Zr and Nb may help to unravel the detailed physics of neutron-source activation, semiconvection, and dredge up in AGB stars of small core mass.

17

Nucleosynthesis in Novae

James W. Truran

17.1 Introduction

It is now commonly accepted that the outbursts of the classical novae are attributable to thermonuclear runaways proceeding in accreted hydrogen-rich envelopes on the white dwarf components of nova binary systems (Gallagher and Starrfield 1978; Truran 1982). For the conditions which are characteristic of this environment, it is expected that nuclear burning will proceed by means of the carbon-nitrogen-oxygen (CNO) cycle hydrogen-burning reactions. The characteristics of these burning sequences—in particular, the limit imposed on the rate of nuclear energy generation at high temperatures by the slower and temperature insensitive positron decays—severely restrict the energetics of the critical stages of these runaways and thereby afford a physical basis for distinguishing between "slow" and "fast" novae (Truran 1982). Hydrodynamic studies of nova eruptions reveal that, following upon runaway, shell hydrogen burning of the residual matter defines a phase of evolution at constant bolometric luminosity (Nariai, Nomoto, and Sugimoto 1980; Prialnik, Shara, and Shaviv 1978, 1979; Sparks, Starrfield, and Truran 1978; Starrfield, Sparks, and Truran 1976; Starrfield, Truran, and Sparks 1978). The evolution during this phase mimics in some respects the evolution of the central stars of planetary nebulae to white dwarfs: the luminosities are consistent with the mass-luminosity relation for degenerate cores (Paczyński 1971; Becker and Iben 1979) and approach the Eddington limit at large core masses. The time scale then depends upon the hydrogen envelope mass available for burning and the effective rate of mass loss via winds (Bath 1978; Ruggles and Bath 1979) or as a consequence of interactions with the binary companion (MacDonald 1980).

It thus appears that nucleosynthesis associated with nova events is limited to the effects of proton-induced reactions which can proceed at the temperatures $\sim 10^8$ to 3×10^8 K achieved in the hydrogen-burning shells. CNO cycle hydrogen burning will of course act to redistribute these nuclei and thereby alter the concentrations of the isotopes of carbon, nitrogen, and oxygen. The observational situation regarding the abundances of these and heavier nuclei in nova ejecta will be reviewed in the next section, and the implications of the observed enrichments of heavy elements will be discussed. In subsequent sections, we review the possible role of novae in the synthesis of ^{7}Li and in the generation of significant abundances of the radioactive nuclei ^{26}Al and ^{22}Na.

It is important to recognize in advance both that the relative contribution of novae to nucleosynthesis is small and that detailed predictions regarding their contributions are quite uncertain. We do not know with confidence either the rate of nova outbursts as a function of time over the history of our Galaxy or the average mass ejected per outburst. The total amount of processed matter contributed by novae to the interstellar medium may be roughly estimated on the following assumptions: (1) the current rate of nova events is 30 yr^{-1}, compatible with the rate observed for Andromeda (Arp 1956); (2) this rate has remained constant over the lifetime $\sim 10^{10}$ yr of the galactic disk; and (3) the average mass ejected per nova event is $\sim 10^{-4}$ $M_\odot$. This implies a total mass $\sim 3 \times 10^7$ $M_\odot$. By comparison, supernovae occurring at a rate ~ 0.03 yr^{-1} and ejecting $\sim M_\odot$ of processed matter per event would yield $\sim 3 \times 10^8$ $M_\odot$ over the same period. Since the mass processed through novae represents only a fraction $\sim 1/300$ of the mass of interstellar matter ($\sim 10^{10}$ $M_\odot$) in our Galaxy, it follows that significant contributions require enrichments, relative to solar system abundances, in nova ejector of factors greater than ~ 300. This would appear to rule out the possibility that novae are important contributors to galactic abundances of the dominant isotopes of carbon and oxygen, and probably nitrogen as well. They may, however, produce important amounts of the rarer isotopes ^{13}C, ^{15}N, and ^{17}O.

17.2 Heavy Element Abundances in Novae

The fact that nova ejecta can exhibit unusual abundance features has been recognized for some time (Payne-Gaposchkin 1957; Mustel 1974). Early curve of growth analyses of novae near visual maximum revealed that CNO nuclei could be enhanced in abundance relative to hydrogen. Williams (1977) argued that the results of such studies are particularly sensitive to the assumed photospheric temperature distribution and are therefore quite uncertain. Most recent investigations have therefore dealt either with the nebular phase of the spectral development of novae in the later stages of decline or, following Williams's suggestion, with the extended remnant shells of ejecta associated with a few old novae.

The abundances determined for nova ejecta from a number of such nebular analyses are collected in table 17.1. Here we present the ratios of the observed abundances to solar system abundances (Cameron 1982), normalized to helium, for elements from carbon to silicon. The row labeled He/H gives the observed helium to hydrogen abundance ratio relative to solar, while $\sum$CNO signifies the total enhancement of the CNO nuclei. Finally, the row labeled Z gives the total mass fraction of the ejecta in the form of heavy elements. Since a determination of the He/H ratio is not available for Nova V693 Corona Austrina 1981, no estimate of the mass fraction of the ejecta in the form of heavy elements is yet possible for this case.

Several important inferences regarding the outbursts of classical novae and the nature of the underlying nova binary systems may be drawn from these data. We note first that, since it is believed that CNO cycle hydrogen burning powers the thermonuclear runaways that define the outbursts of classical novae, some rearrangement of the elemental and isotopic abundance patterns is to be anticipated. In particular, one should expect an elemental enrichment of nitrogen relative to carbon and oxygen just as is found for all the novae included in the table. Overabundances of the rarer isotopes of CNO nuclei, including ^{13}C and ^{15}O, are also expected: the calculated nova models of Starrfield et al. (1978), for example, predict average enrichment factors $\sim 10^2$–10^3 for ^{13}C and ^{15}O in the nova ejecta. An analysis by Sneden and Lambert (1975) of CN bands observed in Nova DQ Her 1934 near maximum light revealed a significant enhancement of ^{13}C and perhaps of ^{15}N as well. Overabundances by factors $\sim$100–1000 are compatible with novae representing a major source of ^{13}C and ^{15}N in the Galaxy.

While elemental and isotopic redistributions among the CNO nuclei themselves may be understood, the extreme overabundances of the elements carbon, nitrogen, and oxygen which characterize the ejecta of several studied novae present a more serious challenge. It may be appropriate here to note that theoretical considerations suggest that significant abundance enrichments are required to explain the rapid development of very fast novae (Truran

Table 17.1
Overabundances in Nova Ejecta Relative to Helium

Nova	C	N	O	ΣCNO	Ne	Na	Mg	Al	Si	He/H	Z
RR Pic 1925[a]	...	8.4	0.35	1.8	3.7	...	...	...	...	2.8	0.039
HR Del 1967[b]	...	9.2	2.5	4.1	0.89	...	...	...	...	3.7	0.077
T Aur 1891[c]	...	32	3.3	8.9	...	...	...	...	...	3.1	0.13
V1500 Cyg 1975[d]	27	100	21	53	21	...	...	...	...	1.0	0.30
V1668Cyg 1978[e]	11	100	15	40	...	...	...	...	...	1.7	0.31
DQ Her 1934[f]	30	400	78	160	...	...	13	63	...	1.0	0.56
V693 CrA 1981[g]	1.3	89	16	32	140	63	13	63	$\approx$1	...	...

a) Williams and Gallagher 1979
b) Tylenda 1978
c) Gallagher et al. 1980
d) Ferland and Shields 1978
e) Stickland et al. 1981
f) Williams et al. 1978
g) Williams et al. 1984

1982). This is a consequence of the characteristics of the CNO cycle hydrogen-burning reaction sequences. The nuclear energy available on a dynamic time scale is constrained by the slower and temperature insensitive weak interactions, in particular the $^{15}O(e^+\nu)^{15}N$ reaction for which $\tau_{1/2} = 176$ s. The energy release on a time scale insufficient to allow appreciable β decay to occur is simply that energy release resulting from the capture of one or two protons on every available CNO nucleus. This gives

$$\epsilon_{nuc} \sim 2 \times 10^{15} \text{ ergs g}^{-1} \left[\frac{N_{CNO}}{N_{CNO}(\theta)} \right]$$

where N_{CNO} is the number density of CNO nuclei in the envelope and $N_{CNO}(\theta)$ is the corresponding solar system value. By comparison, the binding energy program of nova envelope matter is

$$\frac{GM}{R} \approx 2 \times 10^{17} \text{ ergs g}^{-1} \frac{M/M_\odot}{R/5 \times 10^8 \text{cm}}$$

It is thus apparent that a significant increase in the concentration of CNO nuclei is required to ensure that the energy release on a dynamic time scale is compatible with the rapid expulsion of a reasonable fraction of the envelope matter.

These theoretical arguments do not address the question as to why most of the studied novae show CNO abundance enrichments, nor do they identify a mechanism by which these very substantial enrichments might be effected. The situation here is now further complicated by the determination that nuclei heavier than carbon, nitrogen, and oxygen may also be present in significant concentrations in nova ejecta. McLaughlin (1960) was perhaps the first to call attention to the fact that strong neon lines were present in the spectra of some novae. Ferland and Shields (1978) found a substantial neon enrichment for Nova V1500 Cyg 1975. Most recently, Williams et al. (1984) determined a neon enrichment relative to helium of a factor greater than 100 for Nova V693 CrA 1981. An even more important result for Nova Corona Austrina is the finding that the elements sodium, magnesium, and aluminum as well as neon are enhanced in abundance by factors comparable to those of the CNO elements. Similar neon and heavy element enrichments have been reported as well for Nova Aquila 1982 by Snijders, Seaton, and Blades (1982).

Possible sources of the heavy element abundance enrichments observed in novae include mass transfer from the secondary, nuclear transformations accompanying the outburst, and upward mixing of matter from the underlying white dwarfs. Some effects due to mass transfer have been reviewed by D'Antona and Mazzitelli (1980). If, for example, an evolved companion were appropriate, the transfer of matter characterized by distorted CNO abundance patterns might be understood. Williams and Ferguson (1983) argue for such transfer on the basis of observations of CNO emission line strengths in nova systems at quiescence. However, the extreme CNO and heavy element enrichments which are sometimes required seem unlikely to have resulted from mass transfer. Nuclear reactions accompanying nova outbursts also appear not to be capable of explaining the observed abundance anomalies. Significant production of heavy nuclei is not expected to result from CNO cycle hydrogen burning at the temperatures $\sim(150\text{--}350) \times 10^6$ K which are characteristic of nova shells. Breakout of the hydrogen-burning sequences via $^{15}O(\alpha,p)^{19}Ne(p,\gamma)^{20}Na$ can occur at higher temperatures (Wallace and Woosley 1981), yielding neon and heavier nuclei, but these more extreme conditions are compatible with thermonuclear runaways on neutron stars rather than white dwarfs. Helium-burning episodes on white dwarfs, which might serve to provide surface enrichments of carbon and oxygen, are also unlikely to occur in the presence of the lower accretion rates, less than $\sim 10^{-8}~M_\odot~\text{yr}^{-1}$, characteristic of classical nova systems. Weak helium shell flashes on a carbon-oxygen white dwarf leading to core growth and possible envelope contamination are predicted to occur for accretion rates less than $\sim 4 \times 10^{-8}~M_\odot$. For helium accretion rates less than $\sim 10^{-8}~M_\odot~\text{yr}^{-1}$, the shell temperatures are lower and ignition is delayed until the helium shell grows substantially in mass: ignition under extremely degenerate conditions is then expected to lead to violent, supernova-like outbursts (Woosley, Weaver, and Taam 1980; Nomoto 1982).

It would thus appear, by a process of elimination, that mixing must provide the mechanism of envelope enrichment in classical nova systems. Since the underlying white dwarfs are generally expected to be carbon-oxygen white dwarfs, this provides a particularly attractive mechanism. One possibility which remains to be explored is that convective overshooting can effect a substantial contamination of the envelope. Detailed analyses to date have been concerned rather with the implications of shear-induced mixing (Kippenhahn and Thomas 1978; Kutter and Sparks 1984; MacDonald 1983). MacDonald finds, in particular, that shear-induced turbulent mixing between the white dwarf and the accreted material can lead to envelope CNO enrichments, provided the thermal time scale is short compared to a few hundred times the accretion time scale. It should be noted that, for the recent novae (Nova Corona Austrina 1981 and Nova Aquila 1982) which show neon-sodium-magnesium-aluminum enrichments, such mixing up of core matter can explain the abundance peculiarities only if one assumes an underlying oxygen-neon-magnesium (ONeMg) white dwarf. We shall return to this point in our subsequent discussion.

17.3 ^{7}Li Production in Novae

The environment in which most of the ^{7}Li in galactic matter has its origin remains to be identified. This situation is complicated by the fact that there are a number of sources which can make some contribution. While the cosmological conditions compatible with ^{7}Li production are nearly coincident with those for deuterium, detailed calculations of primordial nucleosynthesis indicate that the abundance level of ^{7}Li resulting from this source is low by approximately a factor of 10. This result is very nicely compatible with the low levels of lithium recently determined to characterize unevolved halo stars and old disk stars (Spite and Spite 1982a, b) and suggests a galactic origin for ^{7}Li. Observations confirm that substantial ^{7}Li production can occur, at least as a transient phase, in the evolution of red giants (Wallerstein and Sneden 1982). Cameron and Fowler (1971) first proposed that high ^{7}Li concentrations in red giant envelopes could be realized if nuclei of mass $A = 7$, in the form of ^{7}Be (which is not as readily destroyed by proton interactions), were convectively mixed outward from the hotter shell-burning regions at the base of the envelope. Unfortunately, predictions of the surface enhancement of ^{7}Li and thus of the integrated contributions of red giants to the ^{7}Li contributions in galactic matter are rendered unreliable by uncertainties in theories of convective transport (Sackmann, Smith, and Despain 1974; Iben 1973, 1975).

Significant production of ^{7}Li can also occur as a concomitant of the thermonuclear runaways which define the outbursts of classical novae (Arnould and Norgaard 1975; Starrfield et al. 1978). The mode of ^{7}Li production here is similar to that in red giants: ^{7}Be is carried outward by convection to cooler regions of the envelope on a sufficiently rapid time scale to ensure that it will not be destroyed via ^{7}Be$(p,\gamma)^8$B. The total mass of ^{7}Li ejected is a sensitive function of the conditions achieved in the outburst and may therefore be expected to vary from event to event. Model calculations (Starrfield et al. 1978) predict that ^{7}Li enrichments of factors $\sim 10^2$–10^3 may characterize the ejecta of novae. The critical dependences with regard to ^{7}Li production are on the temperature history of the matter and the convective character of the envelope over the course of the runaway (both of which may be a function of the speed class of the nova); the initial ^{3}He concentration of the envelope matter is also of interest since much of the ^{3}He may be converted readily to ^{7}Be.

The integrated contributions of classical novae to the abundance of ^{7}Li in galactic matter cannot be too reliably estimated due to uncertainties associated with our knowledge of the properties of the progenitors of classical nova systems and of their rate of formation over the past history of the Galaxy. Nevertheless, a statement regarding production of ^{7}Li in nova is possible. The calculated models of Starrfield et al. (1978) predict average enrichment factors for the rare isotopes ^{13}C and ^{15}N which are approximately 10 times that of ^{7}Li. If these relative production ratios are representative of classical nova systems, it must follow that novae can account at best for only $\sim 10\%$ of the ^{7}Li in the Galaxy—when one assumes that ^{13}C and ^{15}N have their origin entirely in nova events. Further detailed calculations of nucleosynthesis coupled to hydrodynamic models of novae are required to clarify the situation for ^{7}Li.

17.4 Synthesis of ^{22}Na and ^{26}Al

The production of the astrophysically important radioactive isotopes ^{22}Na and ^{26}Al can also occur under the explosive hydrogen-burning conditions that are associated with classical nova outbursts. The formation of significant concentrations of mass 22 as ^{22}Na($\tau_{1/2} = 2.6y$) provides a potentially attractive explanation for the highly ^{22}Ne-enriched Ne-E anomaly (Black 1972; Eberhardt et al. 1979; Lewis et al. 1979) identified in meteorites. ^{26}Al is of course important in that the existence of ^{26}Mg excesses in meteoritic matter that correlate with Al/Mg ratios (Lee, Papanastassiou, and Wasserburg 1977 a, b) confirms the presence of live ^{26}Al($\tau_{1/2} = 7.2 \times 10^5 y$) in the early solar system.

Quantitative investigations of ^{22}Na and/or ^{26}Al production in explosive hydrogen burning have been performed by a number of authors (Arnould and Norgaard 1978; Arnould et al. 1980; Vangioni-Flam et al. 1980; Hillebrandt and Thielemann 1982). Predictions of the ^{22}Na and ^{26}Al concentrations in nova ejecta are rendered unreliable by uncertainties associated both with the relevant nuclear reaction cross sections in the energy range of interest and with the temperature history of the ejected matter. We will briefly review the situation for both ^{22}Na and ^{26}Al in the following discussion.

Early studies of ^{22}Na production in novae gave rise to predictions of rather substantial concentrations. If only it was assumed that all neon and sodium, present initially in solar system proportions, were converted into ^{22}Na as a consequence of the neon-sodium-magnesium cycle reactions, an enrichment of mass 22 as ^{22}Na of a factor $\sim$8 was realized. The presence of ^{22}Na at this level in nova ejecta provides a potentially detectable gamma-ray source (Clayton and Hoyle 1974; Leventhal et al. 1977; Truran et al. 1978). The revised rates for the reactions ^{21}Na$(p,\gamma)^{22}$Mg and ^{22}Na$(p,\gamma)^{23}$Mg provided by Wallace and Woosley (1981) are larger than those used in earlier work by some orders of magnitude. These rates ensure a far more rapid rate of ^{22}Na destruction in the temperature range 1–3 $\times$ 10^8 K sampled by the hydrogen-burning shells of novae in outbursts.

Important questions regarding ^{22}Na production in the light of these revised rates include the following. Does the implied level of ^{22}Na production constitute a detectable gamma-ray source? Are the abundances of masses A = 20, 21, and 22 formed in situ consistent with the neon isotopic distribution in the Ne-E component? Is the ratio ^{22}Na/^{23}Na predicted for nova ejecta compatible with the realization of a Ne-E component? The calculations of nucleosynthesis in novae by Hillebrandt and Thielemann (1982) provide useful input to some of these questions. They find, in particular, that the concentrations of ^{20}Ne and ^{21}Ne formed relative to ^{22}Ne + ^{22}Na are too large to explain the meteoritic Ne-E component. On the other hand, the ^{22}Na/^{23}Na production ratio they obtain is quite compatible with the formation of a component of the character of Ne-E as a consequence of ^{22}Na decay in grains formed in the expansion and cooling of nova ejecta.

Interest in the subject of ^{26}Al production in novae has been stimulated both by evidence for the presence of ^{26}Al in primitive solar system matter (Wasserburg and Papanastassiou 1982) and by *HEAO 3* observations of gamma-rays from the decay of ^{26}Al in the interstellar medium (Mahoney et al. 1982). It was originally believed that the source of the ^{26}Al in the early solar system was most likely a supernova. Detailed calculations indicate that the ^{26}Al/^{27}Al production ratio in explosive nucleosynthesis in supernova environments must lie close to 10^{-3} (Truran and Cameron 1978; Arnett and Wefel 1978; Woosley and Weaver 1980). Clayton (1984) has argued that this is too low a production ratio, given the amount of dispersed interstellar ^{26}Al detected by *HEAO 3:* the implied concentration of ^{27}Al in galactic matter is considerably larger than is observed. In contrast, the production ratio ^{26}Al/^{27}Al can be close to unity for matter processed through high temperature hydrogen burning in novae. Hillebrandt and Thielemann (1982) find specifically that ^{26}Al is coproduced with ^{22}Na, with production ratios ^{26}Al/^{27}Al $\sim$ 1–2. They also note that the concentration of ^{26}Al present in nova ejecta is strongly dependent upon the rates of certain critical reactions. Since revised rates are now available for both the ^{25}Mg$(p,\gamma)^{26}$Al reaction (Champagne, Howard, and Parker 1983) and the ^{26}Mg$(p,n)^{26}$Al reaction (Skelton, Kavanagh, and Sargood 1983), further calculations of ^{26}Al production in novae seem appropriate.

17.5 Discussion and Conclusions

In spite of the uncertainties associated with existing hydrodynamic models of the outbursts of classical novae, several general conclusions may be drawn concerning nuclear processes in novae and their contributions to galactic nucleosynthesis.

1. Spectroscopic studies have established that significant enrichments of the elements carbon, nitrogen, and oxygen characterize the ejecta of novae. For the six novae listed in table 17.1 for which abundances of both hydrogen and heavy elements are known, the average enrichments relative to hydrogen are by factors $\sim$25 for carbon and oxygen and a factor $\sim$140 for nitrogen. Let us simply assume for our present purposes that these averages are appropriate to all novae and ask what this would imply. Isotopic information is of course not generally available. The overproduction factors for carbon and oxygen are not sufficient to make novae important sources of ^{12}C and ^{16}O. The situation regarding ^{14}N is less obvious. Williams (1982) has argued on the basis of these abundance data that novae may be important contributors to the ^{14}N abundances of galaxies. The critical factor here would appear to be the abundance of ^{14}N in nova ejecta relative to such rarer nuclear species as ^{13}C, ^{15}N, and ^{17}O: model predictions to date reveal that these less abundant isotopes can be enriched in abundance by larger factors. This question clearly demands further observational and theoretical study.

2. Significant neon abundance enrichments are also found in novae. In the cases of the recent Nova V693 CrA 1981 (and perhaps Nova Aql 1982 as well), this is accompanied by large overabundances of sodium, magnesium, and aluminum. As we have seen, such abundance configurations cannot easily be understood on the basis either of mass transfer or of the hydrogen- or helium-burning sequences which may proceed in nova envelopes. Mixing of heavy elements outward from the core thus seems to have occurred. Such mixing up of core matter cannot explain these abundance peculiarities if the underlying white dwarf is composed of carbon and oxygen: we are thus led to propose that we may be dealing here rather with an ONeMg white dwarf.

Evidence for the existence of such configurations in nova binary systems would have interesting implications. Stellar evolution studies predict that stars in the mass range $\sim$1–9 $M_\odot$ will form CO degenerate cores while those of initial mass $\sim$8–12 $M_\odot$ will evolve further to yield ONeMg degenerate cores (Barkat, Reiss, and Rakavy 1974). Assuming a Salpeter (1955) initial mass spectrum, these mass ranges imply that the ratio of stars with CO cores to those with ONeMg cores is approximately 35. Effects attributable to binary evolution can act to alter the relative frequencies. Law and Ritter (1983) first suggested that ONeMg white dwarfs might be found in cataclysmic variable systems. Iben and Tutukov (1984) derived realization frequencies for systems containing CO and ONeMg white dwarfs, and determined a ratio of systems containing CO dwarfs to those with ONeMg dwarfs which was close to the single star value $\sim$35. The frequency of occurrence of ONeMg white dwarfs, as inferred from observations of nova ejecta, would seem to be much larger than is predicted. This may reflect a selection effect attributable to the fact that the ONeMg white dwarfs will typically have larger masses than the CO white dwarfs formed under such circumstances. For a fixed accretion rate, these more massive systems should therefore experience nova outbursts more frequently, as they required lower accumulated envelope masses to trigger thermonuclear runaway. Further observations are required to establish whether indeed there exists a population of ONeMg white dwarfs in cataclysmic variable systems.

3. Classical novae can produce ^{7}Li in sufficient quantities to perhaps make an important contribution to the abundance of ^{7}Li in the Galaxy. The critical uncertainties here are those associated with the convective history of nova envelope matter following runaway, with estimates of the average

amount of matter ejected in a nova event, and with predictions of the relative enrichments of ^{7}Li and other nuclei in nova ejecta.

4. The calculations of Hillebrandt and Thielemann (1982) indicate that nucleosynthesis in novae can provide a source of the meteoritic Ne-E component. The isotopic ratios ^{20}Ne/^{22}Ne and ^{21}Ne/^{22}Ne formed under these circumstances are too large to explain Ne-E. However, the ·formation of a sufficient amount of mass $A = 22$ as ^{22}Na, with a ^{22}Na/^{23}Na production ratio well in excess of the observed value in the meteoritic material, allows Ne-E to be explained as a decay product of ^{22}Na formed in novae. We should include a cautionary note here that the presence of substantial neon and magnesium enrichments in some novae might strongly influence the production of both ^{22}Na and ^{26}Al.

5. Hillebrandt and Thielemann (1982) have found that ^{26}Al is also synthesized in nova envelopes. The amount formed is strongly dependent on several critical rates of production and destruction, which are themselves very temperature sensitive. They obtain ^{26}Al/^{27}Al production ratios approaching 1 for these H-burning sequences. Clayton (1984) has argued both that such a high production ratio is demanded, if we are to explain the observed gamma-ray line strengths from ^{26}Al without overproducing ^{27}Al in galactic matter, and that novae are capable of the required level of ^{26}Al production. We believe that the envelopes of red giants provide an equally attractive alternative site for ^{26}Al production. The recent experimental study of the ^{25}Mg$(p,\gamma)^{26}$Al reaction by Champagne, Howard, and Parker (1983) led to the finding that the stellar reaction rate at low temperatures is larger by orders of magnitude than had previously been thought. It thus appears that 'a considerable fraction of the ^{25}Mg in the envelopes of asymptotic giant branch stars may be converted to ^{26}Al. Our quantitative estimate, based on the study of nucleosynthesis in thermally pulsing red giant stars by Iben and Truran (1978), indicates that red giants can produce sufficient ^{26}Al to explain the observed gamma-ray line intensities (Mahoney et al. 1982). A production ratio ^{26}Al/^{27}Al $\sim$ 1 may also be expected for this environment, although a proper quantitative study is required to determine the effects of ^{26}Al decay occurring in the envelope matter over the lifetime of the asymptotic giant branch phase. It is important to determine whether novae are indeed required as a source of ^{26}Al, for this imposes useful constraints on nova hydrodynamic models.

Acknowledgments

This research was supported in part by the National Science Foundation under grants AST 80-18198 and AST 83-14415.

References

Arnett, W. D., and Wefel, J. P. 1978, *Ap. J. (Letters)*, **224**, L139.

Arnould, M., and Norgaard, H. 1975, *Astr. Ap.*, **42**, 55.

——. 1978, *Astr. Ap.*, **64**, 195.

Arnould, M.; Norgaard, H.; Thielemann, F.-K.; and Hillebrandt, W. 1980, *Ap. J.*, **237**, 931.

Arp, H. C. 1956, *Astr. J.*, **61**, 15.

Barkat, Z., Reiss, Y., and Rakavy, G. 1974, *Ap. J. (Letters)*, **193**, L21.

Bath, G. T. 1978, *M.N.R.A.S.*, **182**, 35.

Becker, S. A., and Iben, I., Jr. 1979, *Ap. J.*, **232**, 831.

Black, D. C. 1972, *Geochim. Cosmochim. Acta*, **36**, 377.

Cameron, A. G. W. 1982, in *Essays in Nuclear Astrophysics*, ed. C. A. Barnes, D. D. Clayton, and D. N. Schramm, p. 23 (Cambridge: Cambridge University Press).

Cameron, A. G. W., and Fowler, W. A. 1971, *Ap. J.*, **164**, 11.

Champagne, A. E., Howard, A. J., and Parker, P. D. 1983, *Ap. J.*, **269**, 686.

Clayton, D. D. 1984, *Ap. J.*, **280**, 144.

Clayton, D. D., and Hoyle, F. 1974, *Ap. J. (Letters)*, **187**, L101.

D'Antona, F., and Mazzitelli, I. 1980, *Ap. J.*, **238**, 229.

Eberhardt, P., Jungck, M. H. A., Meier, F. O., and Niederer, F. 1979, *Ap. J. (Letters)*, **234**, L169.

Ferland, G. J., and Shields, G. A. 1978, *Ap. J.*, **226**, 172.

Gallagher, J. S., Hege, E. K., Kopriva, D. A., Williams, R. E., and Butcher, H. R. 1980, *Ap. J.*, **237**, 55.

Gallagher, J. S., and Starrfield, S. 1978, *Ann. Rev. Astr. Ap.*, **16**, 171.

Hillebrandt, W., and Thielemann, F.-K. 1982, *Ap. J.*, **255**, 617.

Iben, I., Jr. 1973, *Ap. J.*, **185**, 209.

——. 1975, *Ap. J.*, **196**, 525.

Iben, I., Jr., and Truran, J. W. 1978, *Ap. J.*, **220**, 980.

Iben, I., Jr., and Tutukov, A. V. 1984, *Ap. J. Suppl.*, **54**, 335.

Kippenhahn, R., and Thomas, H.-C. 1978, *Astr. Ap.*, **63**, 265.

Kutter, S. G., and Sparks, W. M. 1984, preprint.

Law, W. Y., and Ritter, H. 1983, *Astr. Ap.*, **123**, 33.

Lee, T., Papanastassiou, D. A., and Wasserburg, G. J. 1977*a*, *Ap. J. (Letters)*, **211**, L107.

———. 1977*b*, *Geochim. Cosmochim. Acta*, **41**, 1473.

Leventhal, M., MacCallum, C., and Watts, A. 1977, *Ap. J.*, **216**, 491.

Lewis, R. S., Alaerts, L., Matsuda, J.-I., and Anders, E. 1979, *Ap. J. (Letters)*, **234**, L165.

MacDonald, J. 1980, *M.N.R.A.S.*, **191**, 933.

———. 1983, *Ap. J.*, **273**, 289.

Mahoney, W. A., Ling, J. C., Jacobson, A. S., and Lingenfelter, R. E. 1982, *Ap. J.*, **262**, 742.

McLaughlin, D. B. 1960, *Ap. J.*, **131**, 739.

Mustel, E. R. 1974, *Vistas in Astronomy*, **16**, 260.

Nariai, K., Nomoto, K., and Sugimoto, D. 1980, *Pub. Astr. Soc. Japan*, **32**, 473.

Nomoto, K. 1982, *Ap. J.*, **253**, 798.

Paczyński, B. 1971, *Acta Astr.*, **21**, 417.

Payne-Gaposchkin, C. 1957, *The Galactic Novae* (Amsterdam: North-Holland).

Prialnik, D., Shara, M. M., and Shaviv, G. 1978, *Astr. Ap.*, **62**, 339.

———. 1979, *Astr. Ap.*, **72**, 192.

Ruggles, C. L. N., and Bath, G. T. 1979, *Astr. Ap.*, **80**, 97.

Sackmann, I.-J., Smith, R. L., and Despain, K. H. 1974, *Ap. J.*, **187**, 555.

Salpeter, E. E. 1955, *Ap. J.*, **121**, 161.

Skelton, R. T., Kavanagh, R. W., and Sargood, D. G. 1983, *Ap. J.*, **271**, 404.

Sneden, C., and Lambert, D. L. 1975, *M.N.R.A.S.*, **170**, 533.

Snijders, M. A. J., Seaton, M. J., and Blades, J. C. 1982, in *Proceedings of the Third European IAU Conference* (Madrid), p. 177.

Sparks, W. M., Starrfield, S., and Truran, J. W. 1978, *Ap. J.*, **220**, 1063.

Spite, F., and Spite, M. 1982*a*, *Astr. Ap.*, **115**, 257.

Spite, M., and Spite, F. 1982*b*, *Nature*, **297**, 483.

Starrfield, S., Sparks, W. M., and Truran, J. W. 1976, in *Structure and Evolution of Close Binary Systems*, ed. P. Eggleton, S. Mitton, and J. Whelan, p. 155 (Dordrecht: Reidel).

Starrfield, S., Truran, J. W., and Sparks, W. M. 1978, *Ap. J.*, **226,** 186.

Starrfield, S., Truran, J. W., Sparks, W. M., and Arnould, M. 1978, *Ap. J.*, **222,** 600.

Stickland, D. J., Penn, C. J., Seaton, M. J., Snijders, M. A. J., and Storey, P. J. 1981, *M.N.R.A.S.*, **197,** 107.

Truran, J. W. 1982, in *Essays in Nuclear Astrophysics*, ed. C. A. Barnes, D. D. Clayton, and D. N. Schramm, p. 467 (Cambridge: Cambridge University Press).

Truran, J. W., and Cameron, A. G. W. 1978, *Ap. J.*, **219,** 226.

Truran, J. W., Starrfield, S. G., and Sparks, W. M. 1978, in *Gamma Ray Spectroscopy in Astrophysics*, ed. T. L. Cline and R. Ramaty (Greenbelt: NASA-79619).

Tylenda, R. 1978, *Acta Astr.*, **28,** 333.

Vangioni-Flam, E., Audouze, J., and Chieze, J.-P. 1980, *Astr. Ap.*, **82,** 234.

Wallace, R. K., and Woosley, S. E. 1981, *Ap. J. Suppl.*, **45,** 389.

Wallerstein, G., and Sneden, C. 1982, *Ap. J.*, **255,** 577.

Wasserburg, G. J., and Papanastassiou, D. A. 1982, in *Essays in Nuclear Astrophysics*, ed. C. A. Barnes, D. D. Clayton, and D. N. Schramm, p. 77 (Cambridge: Cambridge University Press).

Williams, R. E. 1977, in *The Interaction of Variable Stars with Their Environments*, ed. R. Kippenhahn, J. Rahe, and W. Strohmeier, p. 242 (Bamberg: Remeis-Sternwarte).

——. 1982, *Ap. J. (Letters)*, **261,** L77.

Williams, R. E., and Ferguson, D. H. 1983, in *Cataclysmic Variables and Related Objects*, ed. M. Livio and G. Shaviv, p. 97 (Dordrecht: Reidel).

Williams, R. E., and Gallagher, J. S. 1979, *Ap. J.*, **228,** 482.

Williams, R. E., Sparks, W. M., Starrfield, S., Ney, E. P., Truran, J. W., and Wyckoff, S. 1984, preprint.

Williams, R. E., Woolf, N. J., Hege, E. K., Moore, R. L., and Kopriva, D. A. 1978, *Ap. J.*, **224,** 171.

Woosley, S. E., and Weaver, T. A. 1980, *Ap. J.*, **238,** 1017.

Woosley, S. E., Weaver, T. A., and Taam, R. E. 1980, in *Proc. Texas Workshop in Type I Supernovae*, ed. J. C. Wheeler, p. 96 (Austin: University of Texas Press).

Contributors

W. David Arnett
Astronomy & Astrophysics Center
University of Chicago
5640 South Ellis Avenue
Chicago, IL 60637

David R. Branch
Department of Physics & Astronomy
University of Oklahoma
Norman, OK 73019

Alastair G. W. Cameron
Harvard College Observatory
60 Garden Street
Cambridge, MA 02138

Donald D. Clayton
Department of Space Physics & Astronomy
Rice University
Houston, TX 77001

John J. Cowan
Department of Physics & Astronomy
University of Oklahoma
440 West Brooks
Norman, OK 73019

William A. Fowler
W. K. Kellogg Radiation Laboratory 106-38
California Institute of Technology
Pasadena, CA 91125

Wolfgang Hillebrandt
Institut für Astrophysik
Max-Planck-Institut für Physik und Astrophysik
Karl-Schwarzschild-Strasse 1
D-8046 Garching bei München
West Germany

Icko Iben, Jr.
Department of Astronomy
341 Astronomy Building
University of Illinois
Urbana, IL 61801

Franz Käppeler
Kernforschungszentrum Karlsruhe GmbH
Institut für Kernphysik III
Postfach 3640
D-7500 Karlsruhe
West Germany

Ewald Müller
Institut für Astrophysik
Max-Planck-Institut für Physik und Astrophysik
Karl-Schwarzschild-Strasse 1
D-8046 Garching bei München
West Germany

Ken'ichi Nomoto
Department of Earth Science & Astronomy
University of Tokyo
Komaba, Meguro-ku
Tokyo 153
Japan

Claus Rolfs
Institut für Kernphysik
Universität Münster
Domagkstrasse-71
D-4400 Münster
West Germany

David N. Schramm
Astronomy & Astrophysics Center
University of Chicago
5640 South Ellis Avenue
Chicago, IL 60637

Gary Steigman
Bartol Research Foundation
University of Delaware
Newark, DE 19716

Friedrich-K. Thielemann
Institut für Astrophysik
Max-Planck-Institut für Physik und Astrophysik
Karl-Schwarzschild-Strasse 1
D-8046 Garching bei München
West Germany

James W. Truran
Department of Astronomy
341 Astronomy Building
University of Illinois
Urbana, IL 61801

Mark E. Wiedenbeck
Enrico Fermi Institute
University of Chicago
933 East 56th Street
Chicago, IL 60637

R. G. Wolff
Institut für Astrophysik
Max-Planck-Institut für Physik und Astrophysik
Karl-Schwarzschild-Strasse 1
D-8046 Garching bei München
West Germany